THE FIRMWARE HANDBOOK

THE FIRMWARE HANDBOOK

Edited by

Jack Ganssle

ELSEVIER

AMSTERDAM • BOSTON • HEIDELBERG • LONDON
NEW YORK • OXFORD • PARIS • SAN DIEGO
SAN FRANCISCO • SINGAPORE • SYDNEY • TOKYO

Newnes is an imprint of Elsevier

Newnes

Newnes is an imprint of Elsevier
200 Wheeler Road, Burlington, MA 01803, USA
Linacre House, Jordan Hill, Oxford OX2 8DP, UK

 Recognizing the importance of preserving what has been written,
Elsevier prints its books on acid-free paper whenever possible.

Library of Congress Cataloging-in-Publication Data

Ganssle, Jack
        The firmware handbook / by Jack Ganssle.
            p. cm.
        ISBN 0-7506-7606-X
        1. Computer firmware. I. Title.

    QA76.765.G36 2004
    004—dc22   2004040238

British Library Cataloguing-in-Publication Data
A catalogue record for this book is available from the British Library.

For information on all Newnes publications
visit our website at www.newnespress.com

Transferred to Digital Printing in 2010.

Acknowledgements

I'd like to thank all of the authors whose material appears in this volume. It's been a big job, folks, but we're finally done!

Thanks also to Newnes acquisition editor Carol Lewis, and her husband Jack, who pestered me for a year before inventing the book's concept, in a form I felt that made sense for the readers and the authors. Carol started my book-writing career over a decade ago when she solicited *The Art of Programming Embedded Systems*. Thanks for sticking by me over the years.

And especially, thanks to my wife, Marybeth, for her support and encouragement, as well as an awful lot of work reformatting materials and handling the administrivia of the project.

—Jack Ganssle, February 2004, Anchorage Marina, Baltimore, MD

Contents

Contents

Contents

Contents

Preface

Welcome to The Firmware Handbook! This book fills a critical gap in the embedded literature. There's a crying need for a compendium of ideas and concepts, a handbook, a volume that lives on engineers' desks, one they consult for solutions to problems and for insight into forgotten areas. Though the main topic is firmware, the gritty reality of the embedded world is that the code and hardware are codependent. Neither exists in isolation; in no other field of software is there such a deep coupling of the real and the virtual.

Analog designers tout the fun of their profession. Sure, it's a gas to slew an op amp. But those poor souls know nothing of the excitement of making things move, lights blink, gas flow. We embedded developers sequence motors, pump blood, actuate vehicles' brakes, control the horizontal and vertical on TV sets and eject CDs. Who can beat the fuzzy line between firmware and the real world for fun?

The target audience for the book is the practicing firmware developer, those of us who write the code that makes the technology of the 21st century work.

The book is not intended as a tutorial or introduction to writing firmware. Plenty of other volumes exist whose goal is teaching people the essentials of developing embedded systems.

Nor is it an introduction to the essentials of software engineering. Every developer should be familiar with Boehm's COCOMO software estimation model, the tenants of eXtreme Programming, Fagen and Gilb's approaches to software inspections, Humphrey's Personal Software Process, the Software Engineering Institute's Capability Maturity Model, and numerous other well-known and evolving methodologies. Meticulous software engineering is a critical part of success in building any large system, but there's little about it that's unique to the embedded world.

The reader also will find little about RTOSes, TCP/IP, DSP and other similar topics. These are hot topics, critical to an awful lot of embedded applications. Yet each subject needs an entire volume to itself, and many, many such books exist.

The material accompanying this book is available on a companion website:
http://www.elsevierdirect.com/companions/9780750676069

SECTION

I

Basic Hardware

Introduction

The earliest electronic computers were analog machines, really nothing more than a collection of operational amplifiers. "Programs" as we know them did not exist; instead, users created algorithms using arrays of electronic components placed into the amps' feedback loops. Programming literally meant rewiring the machine. Only electrical engineers understood enough of the circuitry to actually use these beasts.

In the 1940s the advent of the stored program digital computer transformed programs from circuits to bits saved on various media… though nothing that would look familiar today! A more subtle benefit was a layer of abstraction between the programmer and the machine. The digital nature of the machines transcended electrical parameters; nothing existed other than a zero or a one. Computing was opened to a huge group of people well beyond just electrical engineers. Programming became its own discipline, with its own skills, none of which required the slightest knowledge of electronics and circuits.

Except in embedded systems. Intel's 1971 invention of the microprocessor brought computers back to their roots. Suddenly computers cost little and were small enough to build into products. Despite cheap CPUs, though, keeping the overall system cost low became the new challenge for designers *and* programmers of these new smart products.

This is the last bastion of computing where the hardware and firmware together form a unified whole. It's often possible to reduce system costs by replacing components with smarter code, or to tune performance problems by taking the opposite tack. Our firmware works intimately with peripherals, sensors and a multitude of real-world phenomena. Though there's been a trend to hire software-only people to build firmware, the very best developers will always be those with a breadth of experience that encompasses the very best software development practices with a w orking knowledge of the electronics.

Every embedded developer should be—no, *must* be—familiar with the information in this chapter. You can't say "well, I'm a firmware engineer, so I don't need to know what a resistor is." You're an *embedded* engineer, tasked with building a system that's more than just code.

A fantastic reference that gives much more insight into every aspect of electronics is *The Art of Electronics* by Horowitz and Hill.

Jack Ganssle writes the monthly Breakpoints column for *Embedded Systems Programming*, the weekly Embedded Pulse rant on embedded.com, and has written four books on embedded systems and one on his sailing fiascos. He started developing embedded systems in the early 70s using the 8008. Since then he's started and sold three electronics companies, including one of the bigger embedded tool businesses. He's developed or managed over 100 embedded products, from deep-sea nav gear to the White House security system. Jack now gives seminars to companies worldwide about better ways to develop embedded systems.

Basic Electronics

Jack Ganssle

DC Circuits

"DC" means *Direct Current*, a fancy term for signals that don't change. Flat lined, like a corpse's EEG or the output from a battery. Your PC's power supply makes DC out of the building's AC (alternating current) mains. All digital circuits require DC power supplies.

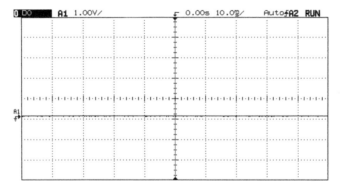

Figure 1-1: A DC signal has a constant, unvarying amplitude.

Voltage and Current

We measure the quantity of electricity using voltage and amperage, but both arise from more fundamental physics. Atoms that have a shortage or surplus of electrons are called *ions*. An ion has a positive or negative charge. Two ions of opposite polarity (one plus, meaning it's missing electrons and the other negative, with one or more extra electrons) attract each other. This attractive force is called the *electromotive force*, commonly known as EMF.

Charge is measured in coulombs, where one coulomb is 6.25×10^{18} electrons (for negative charges) or protons for positive ones.

An ampere is one coulomb flowing past a point for one second. Voltage is the force between two points for which one ampere of current will do one joule of work, a joule per second being one watt.

But few electrical engineers remember these definitions and none actually use them.

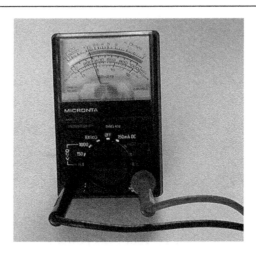

*Figure 1-2: A VOM, even an old-fashioned analog model like this $10
Radio Shack model, measures DC voltage as well or better than a scope.*

An old but still apt analogy uses water flow through a pipe: current would be the amount of water flowing through a pipe per unit time, while voltage is the pressure of the water.

The unit of current is the ampere (amp), though in computers an amp is an awful lot of current. Most digital and analog circuits require much less. Here are the most common nomenclatures:

Name	Abbreviation	# of amps	Where likely found
amp	A	1	Power supplies. Very high performance processors may draw many tens of amps.
milliamp	mA	.001 amp	Logic circuits, processors (tens or hundreds of mA), generic analog circuits.
microamp	μA	10^{-6} amp	Low power logic, low power analog, battery backed RAM.
picoamp	pA	10^{-12} amp	Very sensitive analog inputs.
femtoamp	fA	10^{-15} amp	The cutting edge of low power analog measurements.

Most embedded systems have a far less extreme range of voltages. Typical logic and microprocessor power supplies range from a volt or two to five volts. Analog power supplies rarely exceed plus and minus 15 volts. Some analog signals from sensors might go down to the millivolt (.001 volt) range. Radio receivers can detect microvolt-level signals, but do this using quite sophisticated noise-rejection techniques.

Resistors

As electrons travel through wires, components, or accidentally through a poor soul's body, they encounter *resistance*, which is the tendency of the conductor to limit electron flow. A vacuum is a perfect resistor: no current flows through it. Air's pretty close, but since water is a decent conductor, humidity does allow some electricity to flow in air.

Superconductors are the only materials with zero resistance, a feat achieved through the magic of quantum mechanics at extremely low temperatures, on the order of that of liquid nitrogen and colder. Everything else exhibits some resistance, even the very best wires. Feel the power cord of your 1500 watt ceramic heater—it's warm, indicating some power is lost in the cord due to the wire's resistance.

We measure resistance in ohms; the more ohms, the poorer the conductor. The Greek capital omega (Ω) is the symbol denoting ohms.

Resistance, voltage, and amperage are all related by the most important of all formulas in electrical engineering. Ohm's Law states:

$$E = I \times R$$

where E is voltage in volts, I is current in amps, and R is resistance in ohms. (EEs like to use "E" for volts as it indicates electromotive force).

What does this mean in practice? Feed one amp of current through a one-ohm load and there will be one volt developed across the load. Double the voltage and, if resistance stays the same, the current doubles.

Though all electronic components have resistance, a *resistor* is a device specifically made to reduce conductivity. We use them everywhere. The volume control on a stereo (at least, the non-digital ones) is a resistor whose value changes as you rotate the knob; more resistance reduces the signal and hence the speaker output.

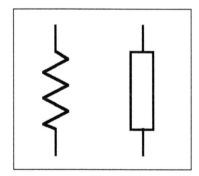

Figure 1-3: The squiggly thing on the left is the standard symbol used by engineers to denote a resistor on their schematics. On the right is the symbol used by engineers in the United Kingdom. As Churchill said, we are two peoples divided by a common language.

Name	Abbreviation	ohms	Where likely found
milliohm	m Ω	.001 ohm	Resistance of wires and other good conductors.
ohm	Ω	1 ohm	Power supplies may have big dropping resistors in the few to tens of ohms range.
hundreds of ohms			In embedded systems it's common to find resistors in the few hundred ohm range used to terminate high speed signals.
kiloohm	k Ω or just k	1000 ohms	Resistors from a half-k to a hundred or more k are found all over every sort of electronic device. "Pullups" are typically a few k to tens of k.
megaohm	M Ω	10^6 ohms	Low signal-level analog circuits.
hundreds of M Ω		10^8++ ohms	Geiger counters and other extremely sensitive apps; rarely seen as resistors of this size are close to the resistance of air.

Table 1-1: Range of values for real-world resistors.

What happens when you connect resistors together? For resistors in series, the total effective resistance is the sum of the values:

$$R_{eff} = R_1 + R_2$$

For two resistors in parallel, the effective resistance is:

$$R_{eff} = \frac{R_1 \times R_2}{R_1 + R_2}$$

(Thus, two identical resistors in parallel are effectively half the resistance of either of them: two 1ks is 500 ohms. Now add a third: that's essentially a 500-ohm resistor in parallel with a 1k, for an effective total of 333 ohms).

The general formula for more than two resistors in parallel is:

$$R_{eff} = \frac{1}{\frac{1}{R_1} + \frac{1}{R_2} + \frac{1}{R_3} + \frac{1}{R_4} + ...}$$

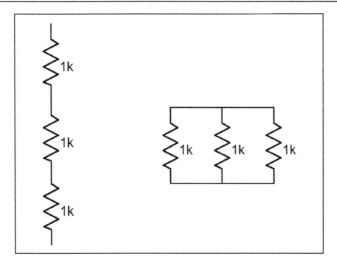

Figure 1-4: The three series resistors on the left are equivalent to a single 3000-ohm part. The three paralleled on the right work out to one 333-ohm device.

Manufacturers use color codes to denote the value of a particular resistor. While at first this may seem unnecessarily arcane, in practice it makes quite a bit of sense. Regardless of orientation, no matter how it is installed on a circuit board, the part's color bands are always visible.

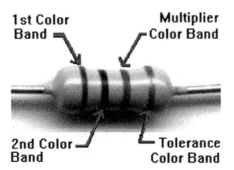

Figure 1-5: This black and white photo masks the resistor's color bands. However, we read them from left to right, the first two designating the integer part of the value, the third band giving the multiplier. A fourth gold (5%) or silver (10%) band indicates the part's tolerance.

Color Band	Value	Multiplier
Black	0	1
Brown	1	10
Red	2	100
Orange	3	1000
Yellow	4	10,000
Green	5	100,000
Blue	6	1,000,000
Violet	7	not used
Gray	8	not used
White	9	not used
Gold (3rd band)		÷10
Silver (3rd band)		÷100

Table 1-2: The resistor color code. Various mnemonic devices designed to help one remember these are no longer politically correct; one acceptable but less memorable alternative is Big Brown Rabbits Often Yield Great Big Vocal Groans When Gingerly Slapped.

The first two bands, reading from the left, give the integer part of the resistor's value. The third is the multiplier. Read the first two band's numerical values and multiply by the scale designated by the third band. For instance: brown black red = 1 (brown) 0 (black) times 100 (red), or 1000 ohms, more commonly referred to as 1k. The following table has more examples.

First band	Second band	Third band	Calculation	Value (ohms)	Commonly called
brown	red	orange	12 x 1000	12,000	12k
red	red	red	22 x 100	2,200	2.2k
orange	orange	yellow	33 x 10,000	330,000	330k
green	blue	red	56 x 100	5,600	5.6k
green	blue	green	56 x 100,000	5,600,000	5.6M
red	red	black	22 x 1	22	22
brown	black	gold	10 ÷10	1	1
blue	gray	red	68 x 100	6,800	6.8k

Table 1-3: Examples showing how to read color bands and compute resistance.

Resistors come in standard values. Novice designers specify parts that do not exist; the experienced engineer knows that, for instance, there's no such thing as a 1.9k resistor. Engineering is a very practical art; one important trait of the good designer is using standard and easily available parts.

Circuits

Electricity always flows in a loop. A battery left disconnected discharges only very slowly since there's no loop, no connection of any sort (other than the non-zero resistance of humid air) between the two terminals. To make a lamp light, connect one lead to each battery terminal; electrons can now run in a loop from the battery's negative terminal, through the lamp, and back into the battery.

There are only two types of circuits: series and parallel. All real designs use combinations of these. A *series circuit* connects loads in a circular string; current flows around through each load in sequence. In a series circuit the current is the same in every load.

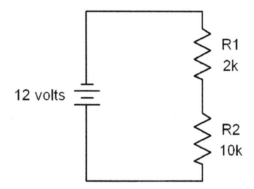

Figure 1-6: In a series circuit the electrons flow through one load and then into another. The current in each resistor is the same; the voltage dropped across each depends on the resistor's value.

It's easy to calculate any parameter of a series circuit. In the diagram above a 12-volt battery powers two series resistors. Ohm's Law tells us that the current flowing through the circuit is the voltage (12 in this case) divided by the resistance (the sum of the two resistors, or 12k). Total current is thus:

$$I = V \div R = (12 \; volts) \div (2000 + 10,000 \; ohms) = 12 \div 12000 = 0.001 \; amp = 1 \; mA$$

(remember that *mA* is the abbreviation for milliamps).

So what's the voltage across either of the resistors? In a series circuit the current is identical in all loads, but the voltage developed across each load is a function of the load's resistance and the current. Again, Ohm's Law holds the secret. The voltage across R1 is the current in the resistor times its resistance, or:

$$V_{R1} = I_{R1} = 0.001 \; amps \times 2000 \; ohms = 2 \; volts$$

Since the battery places 12 volts across the entire resistor string, the voltage dropped on R2 must be 12 – 2, or 10 volts. Don't believe that? Use Mr. Ohm's wonderful equation on R2 to find:

$$V_{R2} = I_{R2} = 0.001 \; amps \times 10,000 \; ohms = 10 \; volts$$

It's easy to extend this to any number of parts wired in series.

Parallel circuits have components wired so both pins connect. Current flows through both parts, though the amount of current depends on the resistance of each leg of the circuit. The voltage, though, on each component is identical.

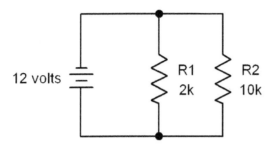

Figure 1-7: R1 and R2 are in parallel, both driven by the 12 volt battery.

We can compute the current in each leg much as we did for the series circuit. In the case above the battery applies 12 volts to both resistors. The current through R1 is:

$$I_{R1} = 12 \ volts \div 2,000 \ ohms = 12 \div 2000 = 0.006 \ amps = 6 \ mA$$

Through R2:

$$I_{R2} = 12 \ volts \div 10,000 \ ohms = 0.0012 \ amps = 1.2 \ mA$$

Real circuits are usually a combination of series and parallel elements. Even in these more complex, more realistic cases it's still very simple to compute anything one wants to know.

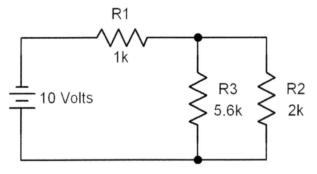

Figure 1-8: A series/parallel circuit.

Let's analyze the circuit shown above. There's only one trick: cleverly combine complicated elements into simpler ones. Let's start by figuring the current flowing out of the battery. It's much too hard to do this calculation till we remember that two resistors in parallel look like a single resistor with a lower value.

Start by figuring the current flowing out of the battery and through R1. We can turn this into a series circuit (in which the current flowing is the same through all of the components) by replacing R3 and R2 by a single resistor with the same effective value as these two paralleled components. That's:

$$R_{EFF} = \frac{R2 \times R3}{R1 + R3} = \frac{5600 \times 2000}{5600 + 2000} = 1474 \ ohms$$

So the circuit is identical to one with two series resistors: R1, still 1k, and R_{EFF} at 1474 ohms. Ohm's Law gives the current flowing out of the battery and through these two resistors:

$$i = \frac{V}{R1 + R_{EFF}} = \frac{10}{1000 + 1474} = 0.004 \ amps = 4 \ mA$$

Ohm's Law remains the font of all wisdom in basic circuit analysis, and readily tells us the voltage dropped across R1:

$$V = iR1 = 0.004 \ amps \times 1000 \ ohms = 4 \ volts$$

Clearly, since the battery provides 10 volts, the voltage across the paralleled pair R2 and R3 is 6 volts.

Power

Power is the product of voltage and current and is expressed in watts. One watt is one volt times one amp. A milliwatt is a thousandth of a watt; a microwatt a millionth.

You can think of power as the total amount of electricity present. A thousand volts sounds like a lot of electricity, but if there's only a microamp available that's a paltry milliwatt—not much power at all.

Power is also current^2 times resistance:

$$P = I^2 \times R$$

Electronic components like resistors and ICs consume a certain amount of volts and amps. An IC doesn't move, make noise, or otherwise release energy (other than exerting a minimal amount of energy in sending signals to other connected devices), so almost all of the energy consumed gets converted to heat. All components have maximum power dissipation ratings; exceed these at your peril.

If a part feels warm it's dissipating a reasonable fraction of a watt. If it's hot but you can keep your finger on it, then it's probably operating within specs, though many analog components want to run cooler. If you pull back, not burned but the heat is too much for your finger, then in most cases (be wary of the wimp factor; some people are more heat sensitive than others) the device is too hot and either needs external cooling (heat sink, fan, etc.), has failed, or your circuit exceeds heat parameters. A burn or near burn, or discoloration of the device, means there's trouble brewing in all but exceptional conditions (e.g., high energy parts like power resistors).

A PC's processor has so many transistors, each losing a bit of heat, that the entire part might consume and eliminate 100+ watts. That's far more than the power required to destroy the chip. Designers expend a huge effort in building heat sinks and fans to transfer the energy in the part to the air.

The role of heat sinks and fans is to remove the heat from the circuits and dump it into the air before the devices burn up. The fact that a part dissipates a lot of energy and wants to run hot is not bad as long as proper thermal design removes the energy from the device before it exceeds its max temp rating.

Figure 1-9: This 10-ohm resistor, with 12 volts applied, draws 833 mA. $P = I^2R$, so it's sucking about 7 watts. Unfortunately, this particular part is rated for ¼ watt max, so is on fire. Few recent college grads have a visceral feel for current, power and heat, so this demo makes their eyes go like saucers.

AC Circuits

AC is short for *alternating current*, which is any signal that's not DC. AC signals vary with time. The mains in your house supply AC electricity in the shape of a sine wave: the voltage varies from a large negative to a large positive voltage 60 times per second (in the USA and Japan) or 50 times (in most of the rest of the world).

AC signals can be either *periodic*, which means they endlessly and boringly repeat forever, or *aperiodic*, the opposite. Static from your FM radio is largely aperiodic as it's quite random. The bit stream on any address or data line from a micro is mostly aperiodic, at least over short times, as it's a complex changing pattern driven by the software.

The rate at which a periodic AC signal varies is called its *frequency*, which is measured in *hertz* (Hz for short). One Hz means the waveform repeats once per second. 1000 Hz is a kHz (kilohertz), a million Hz is the famous MHz by which so many microprocessor clock rates are defined, and a billion Hz is a GHz.

The reciprocal of Hz is *period*. That is, where the frequency in hertz defines the signal's repetition rate, the period is the time it takes for the signal to go through a cycle. Mathematically:

$$\text{Period in seconds} = 1 \div \text{frequency in Hz}$$

Thus, a processor running at 1 GHz has a clock period of 1 nanosecond—one billionth of a second. No kidding. In that brief flash of time even light goes but a bare foot. Though your 1.8 GHz PC may seem slow loading Word®, it's cranking instructions at a mind-boggling rate.

Wavelength relates a signal's period—and thus its frequency—to a physical "size." It's the distance between repeating elements, and is given by:

$$\text{Wavelength in meters} = \frac{c}{frequency} = \frac{300,000,000 \; meters \, / \, second}{frequency \; in \; Hz}$$

where *c* is the speed of light.

An FM radio station at about 100 MHz has a wavelength of 3 meters. AM signals, on the other hand, are around 1 MHz so each sine wave is 300 meters long. A 2.4-GHz cordless phone runs at a wavelength a bit over 10 cm.

As the frequency of an AC signal increases, things get weird. The basic ideas of DC circuits still hold, but need to be extended considerably. Just as relativity builds on Newtonian mechanics to describe fast-moving systems, electronics needs new concepts to properly describe fast AC circuits.

Resistance, in particular, is really a subset of the real nature of electronic circuits. It turns out there are three basic kinds of resistive components; each behaves somewhat differently. We've already looked at resistors; the other two components are capacitors and inductors. Both of these parts exhibit a kind of resistance that varies depending on the frequency of the applied signal; the amount of this "AC resistance" is called reactance.

Capacitors

A capacitor, colloquially called the "cap," is essentially two metal plates separated from each other by a thin insulating material. This insulation, of course, means that a DC signal cannot flow through the cap. It's like an open circuit.

But in the AC world strange things happen. It turns out that AC signals can make it across the gap between the two plates; as the frequency increases the effective resistance of this gap decreases. This resistive effect is called *reactance*; for a capacitor it's termed *capacitive reactance*. There's a formula for everything in electronics; for capacitive reactance it's:

$$X_c = \frac{1}{2\pi fc}$$

where:

X_c = capacitive reactance

f = frequency in Hz

c = capacitance in farads

15

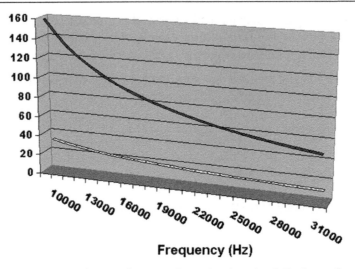

Frequency (Hz)

Figure 1-10: Capacitive reactance of a 0.1 μF cap (top) and a 0.5 μF cap (bottom curve). The vertical axis is reactance in ohms. See how larger caps have lower reactances, and as the frequency increases reactance decreases. In other words, a bigger cap passes AC better than a smaller one, and at higher frequencies all caps pass more AC current. Not shown: at 0 Hz (DC), reactance of all caps is essentially infinite.

Capacitors thus pass only *changing* signals. The current flowing through a cap is:

$$I = \frac{dV}{dt}$$

(If your calculus is rusty or nonexistent, this simply means that the current flow is proportional to the change in voltage over time.)

In other words, the faster the signal changes, the more current flows.

Name	Abbreviation	farads	Where likely found
picofarad	pF	10^{-12} farad	Padding caps on microprocessor crystals, oscillators, analog feedback loops.
microfarad	μF	10^{-6} farad	Decoupling caps on chips are about .01 to .1μF. Low freq decoupling runs about 10μF, big power supply caps might be 1000μF.
farad	F	1 farad	One farad is a huge capacitor and generally does not exist. A few vendors sell "supercaps" that have values up to a few farads but these are unusual. Sometimes used to supply backup power to RAMs when the system is turned off.

Table 1-4: Range of values for real-world capacitors.

In real life there's no such thing as a perfect capacitor. All leak a certain amount of DC and exhibit other more complex behavior. For that reason, there's quite a range of different types of parts.

In most embedded systems you'll see one of two types of capacitors. The first are the polarized ones, devices which have a plus and a minus terminal. Connect one backwards and the part will likely explode!

Polarized devices have large capacitance values: tens to thousands of microfarads. They're most often used in power supplies to remove the AC component from filtered signals. Consider the equation of capacitive reactance: large cap values pass lower frequency signals efficiently. Typical construction today is from a material called "tantalum"; seasoned EEs often call these devices "tantalums." You'll see tantalum caps on PC boards to provide a bit of bulk storage of the power supply.

Smaller caps are made from a variety of materials. These have values from a few picofarads to a fraction of a microfarad. Often used to "decouple" the power supply on a PCB (i.e., to short high frequency switching from power to ground, so the logic signals don't get coupled into the power supply). Most PCBs have dozens or hundreds of these parts scattered around.

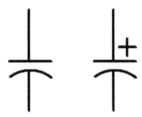

Figure 1-11: Schematic symbols for capacitors. The one on the left is a generic, generally low-valued (under 1 μF) part. On the right the plus sign shows the cap is polarized. Installed backwards, it's likely to explode.

We can wire capacitors in series and in parallel; compute the total effective capacitance using the rules opposite those for resistors. So, for two caps in parallel sum their values to get the effective capacitance. In a series configuration the total effective capacitance is:

$$C_{eff} = \frac{1}{\dfrac{1}{C_1} + \dfrac{1}{C_2} + \dfrac{1}{C_3}\dots}$$

Note that this rule is for figuring the total capacitance of the circuit, and *not* for computing the total reactance. More on that shortly.

One useful characteristic of a capacitor is that it can store a charge. Connect one to a battery or power supply and it will store that voltage. Remove the battery and (for a perfect, lossless part) the capacitor will still hold that voltage. Real parts leak a bit; ones rated at under 1 μF or so discharge rapidly. Larger parts store the charge longer.

Interesting things happen when wiring a cap and a resistor in series. The resistor limits current to the capacitor, causing it to charge slowly. Suppose the circuit shown in the following diagram is dead, no voltage at all applied. Now turn on the switch. Though we've applied a DC signal, the sudden transition from 0 to 5 volts is AC.

Current flows due to the $I = \dfrac{dV}{dt}$ rule; dV is the sudden edge from flipping the switch.

But the input goes from an AC-edge to steady-state DC, so current stops flowing pretty quickly. How fast? That's defined by the circuit's *time constant.*

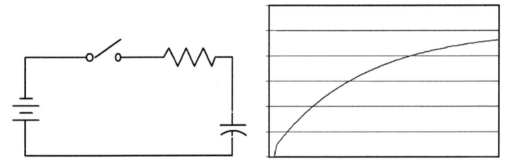

Figure 1-12: Close the switch and the voltage applied to the RC circuit looks like the top curve. The lower graph shows how the capacitor's voltage builds slowly with time, headed asymptotically towards the upper curve.

A resistor and capacitor in series is colloquially called an RC circuit. The graph shows how the voltage across the capacitor increases over time. The time constant of any circuit is pretty well approximated by:

$$t = RC$$

for *R* in ohms, *C* in farads, and *t* in seconds.

This formula tells us that after *RC* seconds the capacitor will be charged to 63.2% of the battery's voltage. After another *RC* seconds another 63.2%, for a total now of 86.5%.

Analog circuits use a lot of RC circuits; in a microprocessor it's still common to see them controlling the CPU's reset input. Apply power to the system and all of the logic comes up, but the RC's time constant keeps reset asserted low for a while, giving the processor time to initialize itself.

The most common use of capacitors in the digital portion of an embedded system is to *decouple* the logic chips' power pins. A medium value part (0.01 to 0.1 μF) is tied between power and ground very close to the power leads on nearly every digital chip. The goal is to keep power supplied to the chips as clean as possible—close to a perfect DC signal.

Why would this be an issue? After all, the system's power supply provides a nearly perfect DC level. It turns out that as a fast logic chip switches between zero and one it can draw immense amounts of power for a short, sub-nanosecond, time. The power supply cannot respond quickly enough to regulate that, and since there's some resistance and reactance between the supply and the chip's pins, what the supply provides and what the chip sees is somewhat different. The decoupling capacitor shorts this very high frequency (i.e., short transient) signal on Vcc to ground. It also provides a tiny bit of localized power storage that helps overcome the instantaneous voltage drop between the power supply and the chip.

Most designs also include a few tantalum bulk storage devices scattered around the PC board, also connected between Vcc and ground. Typically these are 10 to 50 µf each. They are even more effective bulk storage parts to help minimize the voltage drop chips would otherwise see.

You'll often see very small caps (on the order of 20 pF) connected to microprocessor drive crystals. These help the device oscillate reliably.

Analog circuits make many wonderful and complex uses of caps. It's easy to build integrators and differentiators from these parts, as well as analog hold circuits that memorize a signal for a short period of time. Common values in these sorts of applications range from 100 pF to fractions of a microfarad.

Inductors

An inductor is, in a sense, the opposite of a capacitor. Caps block DC but offer diminishing resistance (really, reactance) to AC signals as the frequency increases. An inductor, on the other hand, passes DC with zero resistance (for an idealized part), but the resistance (reactance) increases proportionately to the frequency.

Physically an inductor is a coil of wire, and is often referred to as a *coil*. A simple straight wire exhibits essentially no inductance. Wrap a wire in a loop and it's less friendly to AC signals. Add more loops, or make them smaller, or put a bit of ferrous metal in the loop, and inductance increases. Electromagnets are inductors, as is the field winding in an alternator or motor.

An *iron core* inductor is wound around a slug of metal, which increases the device's inductance substantially.

Inductance is measured in henries (H). *Inductive reactance* is the tendency of an inductor to block AC, and is given by:

$$X_L = 2\pi L f$$

where:

X_L = Inductive reactance

f = frequency in Hz

L = inductance in henries

Clearly, as the frequency goes to zero (DC), reactance does as well.

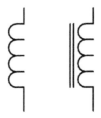

Figure 1-13: Schematic symbols of two inductors. The one on the left is an "air core"; that on the right an "iron core."

Inductors follow the resistor rules for parallel and series combinations: add the value (in henries) when in series, and use the division rule when in parallel.

Inductors are much less common in embedded systems than are capacitors, yet they are occasionally important. The most common use is in switching power supplies. Many datacomm circuits use small inductors (generally millihenries) to match the network being driven.

Power supplies usually have a *transformer* which reduces the AC mains (from the wall) to a lower voltage more appropriate for embedded systems.

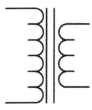

Figure 1-14: The schematic symbol for a transformer.

Transformers are two inductors wrapped around each other, with an iron core. The input AC generates a changing magnetic field, which induces a voltage in the output ("secondary") inductor.

If both inductors have the same number of wire loops, the output voltage is the same as the input. If the secondary has fewer loops, the voltage is less.

Sometimes signals, especially those flowing off a PC board, will have a *ferrite bead* wrapped around the wire. These beads are small cylinders (a few mm long) made of a ferromagnetic material. Like all inductors they help block AC so are used to minimize noise of signal wires.

Active Devices

Resistors, capacitors and inductors are the basic *passive* components, passive meaning "dumb." The parts can't amplify or dramatically change applied signals. By contrast, *active* parts can clip, amplify, distort and otherwise change an applied signal.

The earliest active parts were vacuum *tubes*, called "valves" in the UK.

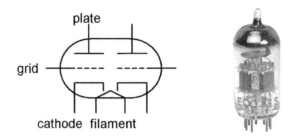

Figure 1-15: On the left, a schematic of a dual triode vacuum tube. The part itself is shown on the right.

Consider the schematic above, which is a single tube that contains two identical active elements, each called a "triode," as each has three terminals. Tubes are easy to understand; let's see how one works.

A *filament* heats the cathode, which emits a stream of electrons. They flow through the grid, a wire mesh, and are attracted to the plate. Electrons are negatively charged, so applying a very small amount of positive voltage to the grid greatly reduces their flow. This is the basis of amplification: a small control signal greatly affects the device's output.

Of course, in the real world tubes are almost unheard of today. When Bardeen, Brattain, and Shockley invented the *transistor* in 1947 they started a revolution that continues today. Tubes are power hogs, bulky and fragile. Transistors—also three-terminal devices that amplify— seem to have no lower limit of size and can run on picowatts.

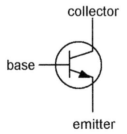

Figure 1-16: The schematic diagram of a bipolar NPN transistor with labeled terminals.

A transistor is made from a single crystal, normally of silicon, into which impurities are doped to change the nature of the material. The tube description showed how it's a voltage controlled device; bipolar transistors are current controlled.

Writers love to describe transistor operation by analogy to water flow, or to the movement of holes and carriers within the silicon crystal. These are at best poor attempts to describe the quantum mechanics involved. Suffice to say that, in the picture above, feeding current into the base allows current to flow between the collector and emitter.

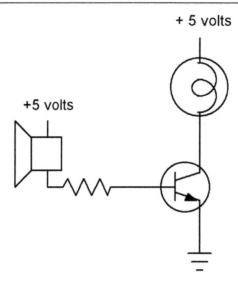

+ 5 volts

+5 volts

Figure 1-17: A very simple amplifier.

And that's about all you need to know to get a sense of how a transistor amplifier works. The circuit above is a trivialized example of one. A microphone—which has a tiny output—drives current into the base of the transistor, which amplifies the signal, causing the lamp to fluctuate in rhythm with the speaker's voice.

A real amplifier might have many cascaded stages, each using a transistor to get a bit of amplification. A radio, for instance, might have to increase the antenna's signal by many millions before it gets to the speakers.

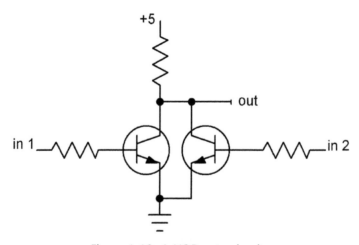

Figure 1-18: A NOR gate circuit.

Transistors are also switches, the basic element of digital circuits. The previous circuit is a simplified—but totally practical—NOR gate. When both inputs are zero, both transistors are off. No current flows from their collectors to emitters, so the output is 5 volts (as supplied by the resistor).

If either input goes to a high level, the associated transistor turns on. This causes a conduction path through the transistor, pulling "out" low. In other words, any input going to a one gives an output of zero. The truth table below illustrates the circuit's behavior.

in1	in2	out
0	0	1
0	1	0
1	0	0
1	1	0

It's equally easy to implement any logic function.

The circuit we just analyzed would work; in the 1960s all "RTL" integrated circuits used exactly this design. But the gain of this approach is very low. If the input dawdles between a zero and a one, so will the output. Modern logic circuits use very high amplification factors, so the output is either a legal zero or one, not some in-between state, no matter what input is applied.

The silicon is a conductor, but a rather lousy one compared to a copper wire. The resistance of the device between the collector and the emitter changes as a function of the input voltage; for this reason active silicon components are called *semiconductors*.

Transistors come in many flavors; the one we just looked at is a bipolar part, characterized by high power consumption but (typically) high speeds. Modern ICs are constructed from MOSFET—Metal Oxide Semiconductor Field Effect Transistor—devices, or variants thereof. A mouthful? You bet. Most folks call these transistors FETs for short.

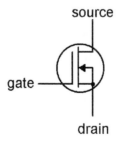

Figure 1-19: The schematic diagram of a MOSFET.

A FET is a strange and wonderful beast. The gate is insulated by a layer of oxide from a silicon channel running between the drain and source. No current flows from the gate to the silicon channel. Yet putting a bias voltage (like a tube, a FET is a voltage device) on the gate creates an electrostatic field that reduces current flow between the other two terminals. Again, *no current flows from the gate*. And when turned on, the source-drain resistance is much lower than in a bipolar transistor. This means the part dissipates little power, a critical concern when putting millions of these transistors on a single IC.

Figure 1-20: The schematic symbol for a diode.

A *diode* is a two-terminal semiconductor that passes current in one direction only. In the picture above, a positive voltage will flow from the left to the right, but not in the reverse direction. Seems a little thing, but it's incredibly useful. The following circuit implements an OR gate without a transistor:

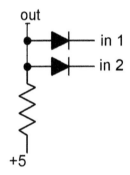

Figure 1-21: A diode OR circuit.

If both inputs are logic one, the output is a one (pulled up to +5 by the resistor). Any input going low will drag the output low as well. Yet the diodes insure that a low-going input doesn't drag the other input down.

Putting it Together—a Power Supply

A power supply is a simple yet common circuit that uses many of the components we've discussed. The input is 110 volts AC (or 220 volts in Europe, 100 in Japan, 240 in the UK). Output might be 5 volts DC for logic circuits. How do we get from high voltage AC input to 5 volts DC?

The first step is to convert the AC mains to a lower voltage AC, as follows:

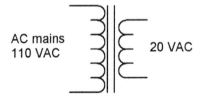

Now let's turn that lower voltage AC into DC. A diode does the trick nicely:

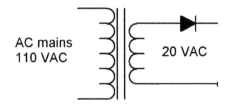

The AC mains are a sine wave, of course. Since the diode conducts in one direction only, its output looks like:

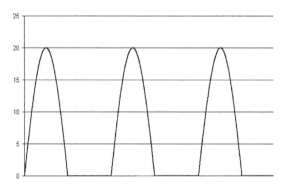

This isn't DC… but the diode has removed all of the negative-going parts of the waveform.

But we've thrown away half the signal; it's wasted. A better circuit uses four diodes arranged in a *bridge* configuration as follows:

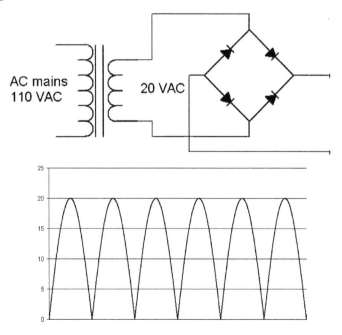

The bridge configuration ensures that two diodes conduct on each half of the AC input, as shown above. It's more efficient, and has the added benefit of doubling the apparent frequency, which will be important when figuring out how to turn this moving signal into a DC level.

The average of this signal is clearly a positive voltage, if only we had a way to create an average value. Turns out that a capacitor does just that:

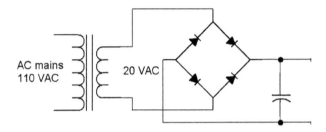

A huge value capacitor filters best—typical values are in the thousands of microfarads.

The output is a pretty decent DC wave, but we're not done yet. The load—the device this circuit will power—will draw varying amounts of current. The diodes and transformer both have resistance. If the load increases, current flow goes up, so the drop across the parts will increase (Ohm's Law tells us $E = IR$, and as I goes up, so does E). Logic circuits are very sensitive to fluctuations in their power, so some form of *regulation* is needed.

A regulator takes varying DC in, and produces a constant DC level out. For example:

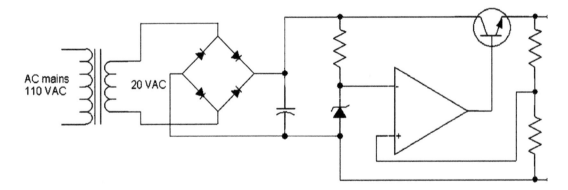

The odd-looking part in the middle is a *zener diode*. The voltage drop across the zener is always constant, so if, for example, this is a 3-volt part, the intersection of the diode and the resistor will *always* be 3 volts.

The regulator's operation is straightforward. The zener's output is a constant voltage. The triangle is a bit of magic—an error amplifier circuit—that compares the zener's constant voltage to the output of the power supply (at the node formed by the two resistors). If the output voltage goes up, the error amplifier applies less bias to the base of the transistor, making it conduct less… and lowering the supply's output. The transistor is key to the circuit; it's sort of like a variable resistor controlled by the error amp.

If, say, 20 volts of unregulated DC goes into the transistor from the bridge and capacitor, and the supply delivers 5 volts to the logic, there's 15 volts dropped across the transistor. If the supply provides even just two amps of current, that's 30 watts (15 volts times two amps) dissipated by that semiconductor—a lot of heat! Careful heatsinking will keep the device from burning up.

The Scope

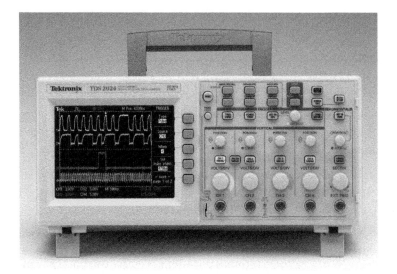

Figure 1-22: A sea of knobs. Don't be intimidated. There's a logical grouping to these. Master them and wow your friends and family. Photo courtesy of Tektronix, Inc.

The oscilloscope (colloquially known as the "scope") is the most basic tool used for trouble-shooting and understanding electronic circuits. Without some understanding of this most critical of all tools, you'll be like a blind person trying to understand color.

The scope has only one function: it displays a graph of the signal or signals you're probing. The horizontal axis is usually time; the vertical is amplitude, a fancy electronics term for voltage.

Controls

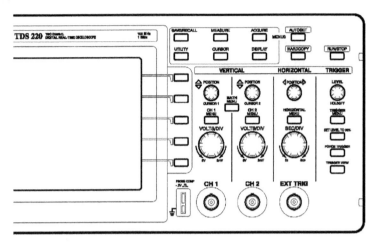

Figure 1-23: Typical oscilloscope front panel. Picture courtesy Tektronix, Inc.

In the above picture note first the two groups of controls labeled "vertical input 1" and "vertical input 2." This is a two-channel scope, by far the most common kind, which allows you to sample and display two different signals at the same time.

The vertical controls are simple. "Position" allows you to move the graphed signal up and down on the screen to the most convenient viewing position. When looking at two signals it allows you to separate them, so they don't overlap confusingly.

"Volts/div" is short for volts-per-division. You'll note the screen is a matrix of 1 cm by 1 cm boxes; each is a "division." If the "volts/div" control is set to 2, then a two volt signal extends over a single division. A five-volt signal will use 2.5 divisions. Set this control so the signal is easy to see. A reasonable setting for TTL (5 volt) logic is 2 volts/div.

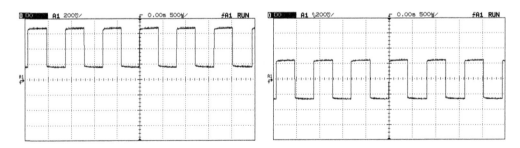

Figure 1-24: The signal is an AC waveform riding on top of a constant DC signal. On the left we're observing it with the scope set to DC coupling; note how the AC component is moved up by the amount of DC (in other words, the total signal is the DC component + the AC). On the right we've changed the coupling control to "AC"; the DC bias is removed and the AC component of the signal rides in the middle of the screen.

The "coupling" control selects "DC"—which means what you see is what you get. That is, the signal goes unmolested into the scope. "AC" feeds the input through a capacitor; since caps cannot pass DC signals, this essentially subtracts DC bias.

The "mode" control lets us look at the signal on either channel, or both simultaneously.

Now check out the horizontal controls. These handle the scope's "time base," so called because the horizontal axis is always the time axis.

The "position" control moves the trace left and right, analogously to the vertical channel's knob of the same name.

"Time/div" sets the horizontal axis' scale. If set to 20 nsec/div, for example, each cm on the screen corresponds to 20 nsec of time. Figure 1-25 shows the same signal displayed using two different time base settings; it's more compressed in the left picture simply because at 2000μsec/div more pulses occur in the one cm division mark.

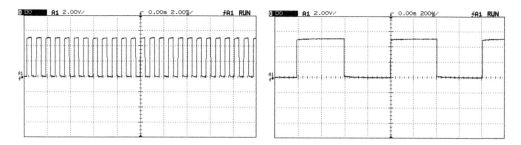

Figure 1-25: The left picture shows a signal with the time base set to 2000 μsec/division; the right is the same signal but now we're sweeping at 200 μsec/division. Though the data is unchanged, the signal looks compressed. Also note that the 5-volt signal extends over 2.5 vertical boxes, since the gain is set to 2 volts/div. The first rule of scoping is to know the horizontal and vertical settings.

The last bank of knobs—those labeled "trigger"—are perhaps the most important of all. Though you see a line on the screen, it's formed by a dot swept across from left to right, repeatedly, at a very high speed. How fast? The dot moves at the speed you've set in the time/div knob. At 1 sec/div the dot takes 10 seconds to traverse the normal 10 cm-wide scope screen. More usual speeds for digital work are in the few microseconds to nanosecond range, so the dot moves faster than any eye can track.

Most of the signals we examine are more or less repetitive: it's pretty much the same old waveform over and over again. The trigger controls tell the scope when to start sweeping the dot across the screen. The alternative—if the dot started on the left side at a random time—would result in a very quickly scrolling screen, which no one could follow.

Twiddling the "trigger level" control sets the voltage at which the dot starts its inexorable left-to-right sweep. Set it to 6 volts and the normal 5-volt logic signal will never get high enough that the dot starts. The screen stays blank. Crank it to zero and the dot runs continuously, unsynchronized to the signal, creating a scrambled mess on the scope screen.

Set trigger-level to 2 volts or so, and as the digital signal traverses from 0 to 5 volts the dot starts scanning, synchronizing now to the signal.

It's most dramatic to learn this control when sampling a sine wave. As you twirl the knob clockwise (from a low trigger voltage to a higher one) the displayed sine wave shifts to the left. That is, the scan starts later and later since the triggering circuit waits for an ever-increasing signal voltage before starting.

"Trigger Menu" calls up a number of trigger selection criterion. Select "trigger on positive edge" and the scope starts sweeping when the signal goes from a low level through the trigger voltage set with the "Trigger Level" knob. "Trigger on negative edge" starts the sweep when the signal falls from a high level through the level.

Every scope today has more features than normal humans can possibly remember, let alone use. Various on-screen menus let you do math on the inputs (add them, etc), store signals that occur once, and much, much more. The instrument is just like a new PC application. Sure, it's nice to read the manual, but don't be afraid to punch a lot of buttons and see what happens. Most functions are pretty intuitive.

Probes

Figure 1-26: Always *connect the probe's ground lead to the system.*

A "probe" connects the scope to your system. Experienced engineers' fingers are permanently bent a bit, warped from too many years holding the scope probe in hand while working on circuit boards. Though electrically the probe is just a wire, in fact there's a bit of electronics magic inside to propagate signals without distortion from your target system to the scope.

So too for any piece of test equipment. The tip of the scope probe is but one of the two connections required between the scope and your target system. A return path is needed, a ground. If there's no ground connection the screen will be nutso, s swirling mass of meaningless scrolling waveforms.

Yet often we'll see engineers probing nonchalantly without an apparent ground connection. Oddly, the waves look fine on the scope. What gives? Where's the return path?

It's in the lab wall. Most electric cords, including the one to the scope and possibly to your target system, have three wires. One is ground. It's pretty common to find the target grounded to the scope via this third wire, going through the wall outlets. Of one thing be sure: even if this ground exists, it's ugly. It's a marginal connection at best, especially when dealing with high-speed logic signals or low level noise-sensitive analog inputs. Never, ever count on it even when all seems well. Every bit of gear in the lab, probably in the entire building, shares this ground. When the Xerox machine on the third floor kicks in, the big inductive spike from the motor starting up will distort the scope signal.

No scope will give decent readings on high-speed digital data unless it is *properly* grounded. I can't count the times technicians have pointed out a clock improperly biased 2 volts above ground, convinced they found the fault in a particular system, only to be bemused and embarrassed when a good scope ground showed the signal in its correct zero to five volt glory. Ground the probe and thus the scope to your target using the little wire that emits from the end of the probe. As circuits get faster, shorten the wire. The very shortest ground lead results in the least signal distortion.

Yet most scope probes come with crummy little lead alligator clips on the ground wire that are impossible to connect to an IC. The frustrated engineer might clip this to a clip lead that has a decent "grabber" end. Those extra 6–12 inches of ground may very well trash the display, showing a waveform that is not representative of reality. It's best to cut the alligator clip off the probe and solder a micrograbber on in its place.

Figure 1-27: Here we probe a complex non-embedded circuit. Note the displayed waveform. A person is an antenna that picks up the 60 Hz hum radiated from the power lines in the walls around us. Some say engineers are particularly sensitive (though not their spouses).

One of the worst mistakes we make is neglecting probes. Crummy probes will turn that wonderful 1-GHz instrument into junk. After watching us hang expensive probes on the floor, mixed in with all sorts of other debris, few bosses are willing to fork over the $150 that Tektronix or Agilent demands. But the $50 alternatives are junk. Buy the best and take good care of them.

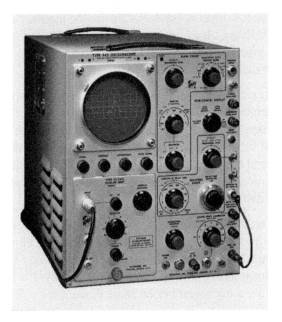

Figure 1-28: Tektronix introduced the 545 scope back in the dark ages; A half-century later many are still going strong. Replace a tube from time to time and these might last forever. About the size of a two drawer file cabinet and weighing almost 100 pounds, they're still favored by Luddites and analog designers.

Logic Circuits

Jack Ganssle

Coding

The unhappy fact that most microprocessor books start with a chapter on coding and number systems reflects the general level of confusion on this, the most fundamental of all computer topics.

Numbers are existential nothings, mere representations of abstract quantitative ideas. We humans have chosen to measure the universe and itemize our bank accounts, so have developed a number of arbitrary ways to count.

All number systems have a *base*, the number of unique identifiers combined to form numbers. The most familiar is decimal, base 10, which uses the ten symbols 0 through 9. Binary is base two and can construct any integer using nothing more than the symbols 0 and 1. Any number system using any base is possible and in fact much work has been done in higher-order systems like base 64—which obviously must make use of a lot of odd symbols to get 64 unique identifiers. Computers mostly use binary, octal (base 8), and hexadecimal (base 16, usually referred to as "hex").

Why binary? Simply because logic circuits are primitive constructs cheaply built in huge quantities. By restricting the electronics to two values only—on and off—we care little if the voltage drifts from 2 to 5. It's possible to build trinary logic, base 3, which uses a 0, 1 and 2. The output of a device in certain ranges represents each of these quantities. But defining three bands means something like: 0 to 1 volt is a zero, 2 to 3 volts a 1, and 4 to 5 a 2. By contrast, binary logic says anything lower than (for TTL logic) 0.8 volts is a zero and anything above 2 a one. That's easy to make cheaply.

Why hex? Newcomers to hexadecimal find the use of letters baffling. Remember that "A" is as meaningless as "5"; both simply represent values. Unfortunately "A" viscerally means something that's not a number to those of us raised to read.

Hex combines four binary digits into a single number. It's compact. "8B" is much easier and less prone to error than "10001011."

Why octal? Base 8 is an aberration created by early programmers afraid of the implications of using letters to denote numbers. It's a grouping of three binary digits to represent the quantities zero through seven. It's less compact than hex, but was well suited to some early mainframe computers that used 36-bit words. 12 octal digits exactly fills one 36-bit word (12 times three bits per digit). Hex doesn't quite divide into 36 bits evenly. Today, though, virtually all computers are 8, 16, 32 or 64 bits, all of which are cleanly divisible by 4, so the octal dinosaur is rarely used.

Decimal	Binary	Octal	Hex	BCD	
00	000000	00	00	0000	0000
01	000001	01	01	0000	0001
02	000010	02	02	0000	0010
03	000011	03	03	0000	0011
04	000100	04	04	0000	0100
05	000101	05	05	0000	0101
06	000110	06	06	0000	0110
07	000111	07	07	0000	0111
08	001000	10	08	0000	1000
09	001001	11	09	0000	1001
10	001010	12	0A	0001	0000
11	001011	13	0B	0001	0001
12	001100	14	0C	0001	0010
13	001101	15	0D	0001	0011
14	001110	16	0E	0001	0100
15	001111	17	0F	0001	0101
16	010000	20	10	0001	0110
17	010001	21	11	0001	0111
18	010010	22	12	0001	1000
19	010011	23	13	0001	1001
20	010100	24	14	0010	0000
21	010101	25	15	0010	0001
22	010110	26	16	0010	0010
23	010111	27	17	0010	0011
24	011000	30	18	0010	0100
25	011001	31	19	0010	0101
26	011010	32	1A	0010	0110
27	011011	33	1B	0010	0111
28	011100	34	1C	0010	1000
29	011101	35	1D	0010	1001
30	011110	36	1E	0011	0000
31	011111	37	1F	0011	0001
32	100000	40	20	0011	0010

Table 2-1: Various coding schemes. BCD is covered a bit later in the text.

To convert from one base to another just remember that the following rule constructs any integer in any number system:

$$Number = ... + C_4 \times b^4 + C_3 \times b^3 + C_2 \times b^2 + C_1 \times b^1 + C_0$$

Each of the C's are coefficients—the digit representing a value, and b is the base. So, the decimal number 123 really is three digits that represent the value:

$$123 = 1 \times 10^2 + 2 \times 10^1 + 3$$

D'oh, right? This pedantic bit of obviousness, though, tells us how to convert any number to base 10. For binary the binary number 10110:

$$10110_2 = 1 \times 2^4 + 0 \times 2^3 + 1 \times 2^2 + 1 \times 2^1 + 0 \times 2^0$$
$$= 22_{10}$$

A1C in hex is:

$$A1C_{16} = A \times 16^2 + 1 \times 16^1 + C \times 16^0$$
$$= 10 \times 16^2 + 1 \times 16^1 + 12 \times 16^0$$
$$= 2588_{10}$$

Converting from decimal to another base is a bit more work. First, you need a cheat sheet, one that most developers quickly memorize for binary and hex, as follows:

Decimal	Binary	Hex
1	2^0	16^0
2	2^1	
4	2^2	
8	2^3	
16	2^4	16^1
32	2^5	
64	2^6	
128	2^7	
256	2^8	16^2
512	2^9	
1024	2^{10}	
2048	2^{11}	
4096	2^{12}	16^3
8192	2^{13}	
16384	2^{14}	
32768	2^{15}	
65536	2^{16}	16^4

To convert 1234 decimal to hex, for instance, use the table to find the largest even power of 16 that goes into the number (in this case 16^2, or 256 base 10). Then see how many times you can subtract this number without the result going negative. In this case, we can take 256 from 1234 four times. The first digit of the result is thus 4.

First digit = 4. Remainder = 1234 − 4 * 256 = 210

Now, how many times can we subtract 16^1 from the remainder (210) without going negative? The answer is 13, or D in hex.

Second digit = D. Remainder = 210 − 13*16= 2

Following the same algorithm for the 16^0 placeholder, we see a final result of 4D2.

Another example: Convert 41007 decimal to hex:

16^3 goes into 41007 10 times before the remainder goes negative, so the first digit is 10 (A in hex).

The remainder is: 41007 − 10 * 16^3 = 47

16^2 can not go into 47 without going negative. The second digit is therefore 0.

16^1 goes into 47 twice. The next digit is 2.

Remainder = 47 − 2 * 16^1 = 15

The final digit is 15 (F in hex).

Final result: A02F

BCD

BCD stands for Binary Coded Decimal. The BCD representation of a number is given in groups of four bits; each group expresses one decimal digit. The normal base 2 binary codes map to the ten digits 0 through 9. Just as the decimal number 10 requires two digits, its equivalent in BCD uses two groups of four bits: 0001 0000. Each group maps to one of the decimal digits.

It's terribly inefficient, as the codes from 1010 to 1111 are never used. Yet BCD matches the base 10 way we count. It's often used in creating displays that show numerics.

Combinatorial Logic

Combinatorial logic is that whose state always reflects the inputs. There's no memory; past events have no impact on present outputs.

An adder is a typical combinatorial device: the output is always just the sum of the inputs. No more, no less.

The easiest way to understand how any combinatorial circuit works—be it a single component or a hundred interconnected ICs—is via a *truth table*, a matrix that defines every possible combination of inputs and outputs. We know, for example, that a wire's output is always the same as its input, as reflected in its table:

In	Out
0	0
1	1

Gates are the basic building blocks of combinatorial circuits. Though there's no limit to the varieties available, most are derived from AND, OR and NOT gates.

NOT Gate

The simplest of all gates inverts the input. It's the opposite of a wire, as shown by the truth table:

In	Out
0	1
1	0

The Boolean expression is a bar over a signal: the NOT of an input A is $\overline{A}$. Any expression can have an inverse, $\overline{A+B}$ is the NOT of $A+B$.

The schematic symbol is:

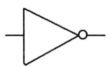

Note the circle on the device's output node. By convention a circle always means inversion. Without it, this symbol would be a buffer: a device that performs no logic function at all (rather like a piece of wire, though it does boost the signal's current). On a schematic, any circle appended to a gate means invert the signal.

AND and NAND Gates

An AND gate combines two or more inputs into a single output, producing a one if *all* of the inputs are ones. If any input is zero, the output will be too.

input1	input2	output
0	0	0
0	1	0
1	0	0
1	1	1

The AND of inputs A and B is expressed as: output = AB

On schematics, a two input AND looks like:

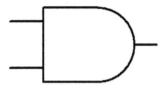

NAND is short for NOT-AND, meaning the output is zero when all inputs are one. It's an AND with an inverter on the output. So the NAND of inputs A and B is: output = $\overline{AB}$.

Schematically a circle shows the inversion:

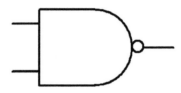

The NAND truth table is:

input1	input2	output
0	0	1
0	1	1
1	0	1
1	1	0

The AND and NAND gates we've looked at all have two inputs. Though these are very common, there's no reason not to use devices with three, four, or more inputs. Here's the symbol for a 13-input NAND gate... its output is zero only when *all* inputs are one:

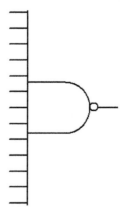

OR and NOR Gates

An OR gate's output is true if *any* input is a one. That is, it's zero only if every input is zero.

input1	input2	output
0	0	0
0	1	1
1	0	1
1	1	1

The OR of inputs A and B is: output = $A+B$

Schematically:

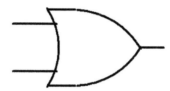

NOR means NOT-OR, and produces outputs opposite that of OR gates:

input1	input2	output
0	0	1
0	1	0
1	0	0
1	1	0

The NOR equation is: output = $\overline{A+B}$. The gate looks like:

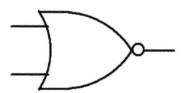

XOR

XOR is short for Exclusive-OR. Often used in error correction circuits, its output goes true if one of the inputs, but not both, is true. Another way of looking at it is the XOR produces a true if the inputs are different.

input1	input2	output
0	0	0
0	1	1
1	0	1
1	1	0

The exclusive OR of A and B is: output $= A \oplus B$

The XOR gate schematic symbol is:

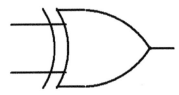

Circuits

Sometimes combinatorial circuits look frighteningly complex, yielding great job security for the designer. They're not. All can be reduced to a truth table that completely describes how each input affects the output(s). Though there are several analysis techniques, truth tables are usually the clearest and easiest to understand.

A proper truth table lists every possible input and output of the circuit. It's trivial to produce a circuit from the complete table. One approach is to ignore any row in the table for which the output is a zero. Instead, write an equation that describes each row with a true output, and then OR these.

Consider the XOR gate previously described. The truth table shows true outputs only when both inputs are different. The Boolean equivalent of this statement, assuming A and B are the inputs, is:

$$\text{XOR} = \overline{A}B + A\overline{B}$$

The circuit is just as simple:

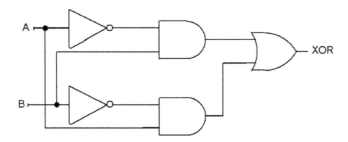

Note that an AND gate combines inputs A and B into AB; another combines the inversions of A and B. An OR gate combines the two product terms into the exclusive OR.

How about something that might seem harder? Let's build an adder, a device that computes the sum of two 16 bit binary numbers.

We could create a monster truth table of 32 inputs, but that's as crazy as the programmer who eschews subroutines in favor of a single, huge, monolithic `main()` function. Instead, realize that each of the 16 outputs (A_0 to A_{15}) is merely the sum of the two single-bit inputs, plus the carry from the previous stage. A 16-bit adder is really nothing more than 16 single-bit addition circuits. Each of those has a truth table as follows:

A_n	B_n	$CARRY_{in}$	SUM_n	$CARRY_{out}$
0	0	0	0	0
0	1	0	1	0
1	0	0	1	0
1	1	0	0	1
0	0	1	1	0
0	1	1	0	1
1	0	1	0	1
1	1	1	1	1

The two outputs (the sum plus a carry bit) are the sum of the A and B inputs, plus the carry out from the previous stage. (The very first stage, for A_0 and B_0, has $CARRY_{in}$ connected to zero).

The one bit adder has two outputs: sum and carry. Treat them independently; we'll have a circuit for each.

The trick to building combinatorial circuits is to minimize the amount of logic needed by not implementing terms that have no effect. In the truth table above we're only really interested in combinations of inputs that result in an output of 1… since any other combination results in zero, by default.

For each truth table row which has a one for the output, write a Boolean term, and then OR each of these as follows:

$$SUM_n = \overline{A_n}B_n\overline{CARRY_{in}} + A_n\overline{B_n}\,\overline{CARRY_{in}} + \overline{A_n}\,\overline{B_n}CARRY_{in} + A_nB_nCARRY_{in}$$

$$CARRY_{out} = A_nB_n\overline{CARRY_{in}} + \overline{A_n}B_nCARRY_{in} + A_n\overline{B_n}CARRY_{in} + A_nB_nCARRY_{in}$$

Each output is a different circuit, sharing only inputs. This implementation could be simplified—note that both outputs share an identical last product term—but in this example we'll pedantically leave it unminimized.

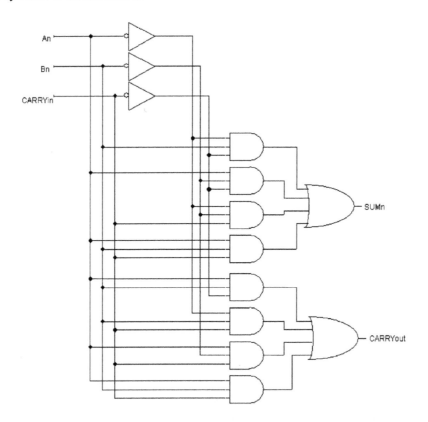

The drawing looks intricate, but does nothing more than the two simple equations above. Drawing the entire circuit for two 16-bit input numbers would cover a rather large sheet of paper, yet is merely a vast amount of duplicated simplicity.

And so, just as programmers manipulate the very smallest of things (ones and zeroes) in massive quantities to create applications, digital designers use gates and other brain-dead bits of minimalism to build entire computers.

This is the essence of all combinatorial design. Savvy engineers will work hard to reduce the complexity of a circuit and therefore reduce parts count, by noticing repetitive patterns, using truth tables, DeMorgan's theorem, and other tools. Sometimes it's hard to figure out how a circuit works, because we're viewing the result of lots of work done to minimize part count. It's analogous to trying to make sense of an object file, without the source. Possible, but tedious.

Tristate Devices

Though all practical digital circuits are binary, there are occasions when it's useful to have a state other than a zero or one. Consider busses: a dozen interconnected RAM chips, for example, all use the same data bus. Yet if more than one tries to drive that bus at a time, the result is babble, chaos. Bus circuits expect each connected component to behave itself, talking only when all other components are silent.

But what does the device do when it is supposed to be quiet? Driving a one or zero is a seriously Bad Thing as either will scramble any other device's attempt to talk. Yet ones and zeroes are the only legitimate binary codes.

Enter the tristate. This is a non-binary state when a bus-connected device is physically turned off. It's driving neither a zero nor a one... rather, the output floats, electrically disconnected from the rest of the circuit.

Bus devices like memories have a control pin, usually named "Output Enable" (OE for short) that, when unasserted, puts the component's output pins into a tristate condition, floating them free of the circuit.

Sequential Logic

The output of sequential logic reflects both the inputs *and the previous state of the circuit*. That is, it remembers the past and incorporates history into the present. A counter whose output is currently 101, for instance, remembers that state to know the next value must be 110.

Sequential circuits are always managed by one or more clocks. A *clock* is a square wave (or at least one that's squarish) that repeats at a fixed frequency. Every sequential circuit is idle until the clock transitions; then, for a moment, everything changes. Counters count. Timers tick. UARTs squirt a serial bit. The clock sequences a series of changes; the circuit goes idle after clock changes to allow signals to settle out.

The clock in a computer ensures that every operation has time to complete correctly. It takes time to access memory, for example. A 50 nsec RAM needs 50 nsec to retrieve data after being instructed as to which location to access. The system clock paces operations so the data has time to appear.

Just as gates are the basic units of combinatorial circuits, flip-flops form all sequential logic. A flip-flop (aka "flop" or "bistable") changes its output based on one or more inputs, after the supplied clock transitions. Databooks show a veritable zoo of varieties. The simplest is the set-reset flop (SR for short), which looks like:

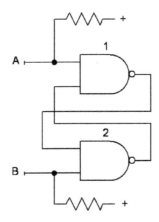

Figure 2-1: SR Flip-flop.

To understand how this works, pretend input A is a zero. Leave B open. It's pulled to a one by the resistor. With A low NAND gate 1 must go to a one, supplying a one to the input gate 2, which therefore, since B is high, must go low. Remove input A and gate 2 still drives a zero into gate 1, keeping output 1 high. The flop has remembered that A was low for a while. Now momentarily drive B low. Like a cat chasing its tail the pattern reverses. Output 1 goes high and 2 low.

What happens if we drive A and B low, and release them at the same time? No one knows. Don't do that.

Flip flops are *latches*, devices that store information. A RAM is essentially an array of many latches.

Possibly the most common of all sequential components is the D flip-flop. As the following drawing shows it has two inputs and one output. The value at the *D* input is transferred to the *Q* output when clock transitions. Change the *D* input and nothing happens till clock comes again.

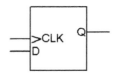

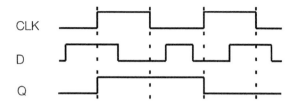

Figure 2-2: Note that the output of the D flip flop changes only on the leading edge of clock.

(Some versions also have set and clear inputs that drive Q to a high or low regardless of the clock. It's not unusual to see a $\overline{Q}$ output as well, which is the inversion of Q.)

Clearly the D flop is a latch (also known as a *register*). String 8 of these together, tied to one common clock, and you've created a byte-wide latch, a parallel output port.

Another common, though often misunderstood, synchronous device is the JK flip flop, named for its inventor (John Kardash). Instead of a single data input (D on the D flip flop), there are two, named J and K. Like the D, nothing happens unless the clock transitions.

But when the clock changes, if J is held low, then the Q output goes to a zero (it follows J). If J is one and K zero, Q also follows J, going to a one.

But if both J and K are one, then the output toggles—it alternates between zero and one every time a clock comes along.

The JK flip flop can form all sorts of circuits, like the following counter:

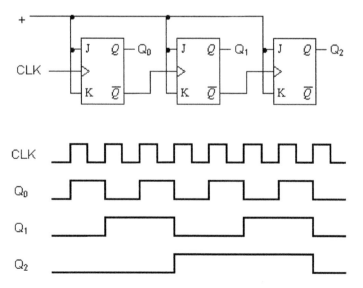

Every clock transition causes the three outputs to change state, counting through the binary numbers. Notice how the clock paces the circuit's operation; it keeps things happening at the rate the designer desires.

Counters are everywhere in computer systems; the program counter sequences fetches from memory. Timer peripherals and real time clocks are counters.

The example above is a *ripple counter*, so called because binary pattern ripples through each flip flop. That's relatively slow. Worse, after the clock transitions it takes some time for the count to stabilize; there's a short time when the data hasn't settled. Synchronous counters are more complex, but switch rapidly with virtually no settle time.

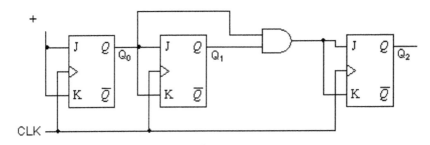

Figure 2-3: A three bit synchronous counter.

Cascading JK flip flops in a different manner creates a shift register. The input bits march (shift) through each stage of the register as the clock operates.

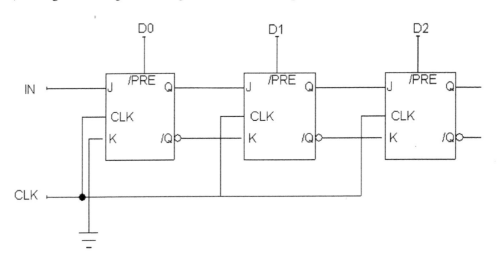

Figure 2-4: A three bit shift register.

Putting a lot of what we've covered together, let's build a simplified UART and driver for an RS-232 device (see following schematic). This is the output part of the system only; considerably more logic is needed to receive serial data. And it doesn't show the start and stop bits. But this drawing shows the use of a counter, a shift register, and discrete parts.

RS-232 data is slow by any standard. Normal microprocessor clocks are far too fast for, say, 9600 baud operation. The leftmost chip is an 8 stage counter. It divides the input frequency by 256. So a 2.5 MHz clock, often found in slower embedded systems, divided as shown, provides 9600 Hz to the shift register.

The register is one that can be parallel-loaded; when the computer asserts the LOAD signal, the CPU's data bus is preset into the 8 stage shift register IC (middle of the drawing). The clock makes this 8 bit parallel word, loaded into the shift register, march out to the QH output, one bit at a time. A simple transistor amplifier translates the logic's 5 volt levels to 12 volts for the RS-232 device.

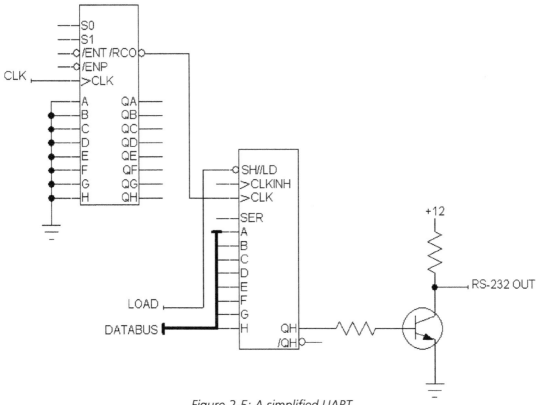

Figure 2-5: A simplified UART.

Logic Wrap-up

Modern embedded systems do make use of all of these sorts of components. However, most designers use integrated circuits that embody complex functions instead of designing with lots of gates and flops. A typical IC might be an 8-bit synchronous counter, or a 4-bit arithmetic-logic unit (that does addition, subtraction, shifting and more).

Yet you'll see gates and flops used as "glue" logic, parts needed to interface big complex ICs together.

Hardware Design Tips

Diagnostics

In the non-embedded world, a favorite debugging trick is to seed print statements into the code. These tell the programmer if the execution stream ever got to the point of the print. But firmware people rarely have this option. So, add a handful of unassigned parallel I/O bits. The firmware people desperately need these as a cheap way to instrument their code. Seeding I/O instructions into the code that drives these outputs is a simple and fast way to see what the program is doing.

Developers can assert a bit when entering a routine or ISR, then drive it low when exiting. A scope or logic analyzer then immediately shows the code snippet's execution time.

Another trick is to cycle an output bit high when the system is busy, and low when idle. Connect a voltmeter to the pin, one of the old fashioned units with an analog needle. The meter will integrate the binary pulse stream, so the displayed voltage will be proportional to system loading.

If space and costs permit, include an entire 8-bit register connected to a row of 0.1 inch spaced vias or headers. Software state-machines can output their current "state" to this port. A logic analyzer captures the data and shows all of the sequencing, with nearly zero impact on the code's execution time.

At least one LED is needed to signal the developer—and perhaps even customers—that the system is alive and working. It's a confidence indicator driven by a low-priority task or idle loop, which shows the system is alive and not stuck somewhere in an infinite loop. A lot of embedded systems have no user interface; a blinking LED can be a simple "system OK" indication.

Highly integrated CPUs now offer a lot of on-chip peripherals, sometimes more than we need in a particular system. If there's an extra UART, connect the pins to an RS-232 level shifting chip (e.g., MAX232A or similar). There's no need to actually load the chip onto the board except for prototyping. The firmware developers may find themselves in a corner where their tools just aren't adequate, and will then want to add a software monitor (see www.simtel.com) to the code. The RS-232 port makes this possible and easy.

If PCB real estate is so limited that there's no room for the level shifter, then at least bring Tx, Rx, and ground to accessible vias so it's possible to suspend a MAX232 on green wires above the circuit board.

(Attention, developers: if you do use this port, don't be in such a panic to implement the monitor that you implement the RS-232 drivers with polled I/O. Take the time to create decent interrupt-driven code. In my experience, polled I/O on a monitor leads to missed characters, an unreliable tool, and massive frustration.)

Bring the reset line to a switch or jumper, so engineers can assert the signal independently of the normal power-up reset. Power-up problems can sometimes be isolated by connecting reset to a pulse generator, creating a repeatable scenario that's easy to study with an oscilloscope.

Connecting Tools

Orient the CPU chip so it's possible to connect an emulator, if you're using one. Sometimes the target board is so buried inside of a cabinet that access is limited at best. Most emulator pods have form factors that favor a particular direction of insertion.

Watch out for vertical clearance, too! A pod stacked atop a large SMT adaptor might need 4 to 6 inches of space above the board. Be sure there's nothing over the top of the board that will interfere with the pod.

Don't use a "clip-on" adaptor on a SMT package. They are just not reliable (the one exception is PLCC packages, which have a large lead pitch). A butterfly waving its wings in Brazil creates enough air movement to topple the thing over. Better, remove the CPU and install a soldered-down adaptor. The PCB will be a prototype forever, but at least it will be a reliable prototype.

Leave margin in the system's timing. If every nanosecond is accounted for, no emulator will work reliably. An extra 5 nsec or so in the read and write cycle—and especially in wait state circuits—does not impact most designs.

If your processor has a BDM or JTAG debug port, be sure to add the appropriate connector on the PCB. Even if you're planning to use a full-blown emulator or some other development tool, at least add PCB pads and wiring for the BDM connector. The connector's cost approaches zero, and may save a project suffering from tool-woes.

A logic analyzer is a fantastic debugging tool, yet is always a source of tremendous frustration. By the time you've finished connecting 100 clip leads, the first 50 have popped off. There's a better solution: surround your CPU with AMP's Mictor connectors. These are high-density, controlled impedance parts that can propagate the system's address, data and control busses off-board. Both Tektronix and Agilent support the Mictor. Both companies sell cables that lead directly from the logic analyzer to a Mictor. No clip leads, no need to make custom cables, and a guaranteed reliable connection in just seconds. Remove the connectors from production versions of the board, or just leave the PCB pads without loading the parts.

Some signals are especially prone to distortion when we connect tools. ALE (address latch enable), also known as AS (address strobe) on Motorola parts, distinguishes address from data on multiplexed busses. The tiniest bit of noise induced from an emulator or even a probe on this signal will cause the system to crash. Ditto for any edge-triggered interrupt input (like

NMI on many CPUs). Terminate these signals with a twin-resistor network. Though your design may be perfect without the terminations, connecting tools and probing signals may corrupt the signals.

Add test points! Unless its ground connection is very short, a scope cannot accurately display the high-speed signals endemic to our modern designs. In the good old days it was easy to solder a bit of wire to a logic device's pins to create an instant ground connection. With SMT this is either difficult or impossible, so distribute plenty of accessible ground points around the board.

Other signals we'll probe a lot, and which must be accessible, include clock, read, write, and all interrupt inputs. Make sure these either have test points, or each has a via of sufficient size that a developer can solder a wire (usually a resistor lead) to the signal.

Do add a Vcc test point. Logic probes are old but still very useful tools. Most need a power connection.

Other Thoughts

Make all output ports readable. This is especially true for control registers in ASICs as there's no way to probe these.

Be careful with bit ordering. If reading from an A/D, for instance, a bad design that flips bit 7 to input bit zero, 6 to 1, etc, is a nightmare. Sure, the firmware folks can write code to fix the mix-up, but most processors aren't good at this. The code will be slow and ugly.

Use many narrow I/O ports rather than a few wide ones. When a single port controls 3 LEDs, two interrupt masks, and a stepper motor, changing any output means managing every output. The code becomes a convoluted mess of ANDs/ORs. Any small hardware change requires a lot of software tuning. Wide ports do minimize part counts when implemented using discrete logic; but inside a PLD or FPGA there's no cost advantage.

Avoid tying unused digital inputs directly to Vcc. In the olden days this practice was verboten, since 74LS inputs were more susceptible to transients than the Vcc pin. All unused inputs went to Vcc via resistor pull-ups. That's no longer needed with logic devices, but is still a good practice. It's much easier to probe and change a node that's not hardwired to power.

However, if you must connect power directly to these unused inputs, be very careful with the PCB layout. Don't run power through a pin; that is, don't use the pin as a convenient way to get the supply to the other pins, or to the other side of the board. It's much better to carefully run all power and ground connections to input signals as tracks on the PCB's outside layers, so they are visible when the IC is soldered in place. Then developers can easily cut the tracks with an X-Acto knife and make changes.

Pullup resistors bring their own challenges. Many debugging tools have their own pull-ups which can bias nodes oddly. It's best to use lower values rather than the high ones permitted by CMOS (say 10k instead of 100k).

PCB silkscreens are oft-neglected debugging aids. Label switches and jumpers. Always denote pin 1 as there's no standard pin one position in the SMT world. And add tick-marks every 5 or 10 pins around big SMT packages, and indicate if pin numbers increase in a CW or CCW direction. Otherwise, finding pin 139 is a nightmare, especially for bifocal-wearing developers suffering from caffeine-induced tremors.

Key connectors so there's no guessing about which way the cable is supposed to go.

Please add comments to your schematic diagrams! For all off-page routes, indicate what page the route goes to. Don't hide the pin numbers associated with power and ground—explicitly label these.

When the design is complete, check every input to every device and make absolutely sure each is connected to something—even if it's not used. I have seen hundreds of systems fail in the field because an unused input drifted to an asserted state. You may expect the software folks to mask these off in the code, but that's not always possible, and even when it is, it's often forgotten.

Try to avoid hardware state machines. They're hard to debug and are often quite closely coupled to the firmware, making that, too, debug-unfriendly. It's easier to implement these completely in the code. Tools (e.g., VisualState from IAR) can automatically generate the state machine code.

Summary

Embedded systems are a blend of hardware and software. Each must complement the other. Hardware people can make the firmware easier to implement.

Many of the suggestions above will make a system easier to debug. Remember that a good design works; a *great* design is also one that's easy to debug.

Designs

Introduction

Jack Ganssle

I talk to engineers all over the world, and hear the same story from so many. They know learning new development techniques would help them get better code out faster, but are in too much of a panic to learn new methods.

Like sailors on a sinking ship, they are too busy bailing to fix the leak. The water slowly rises, so they bail ever more frantically. Sooner or later they're going down, but working faster and harder staves off the inevitable end for just a while longer.

An old cartoon shows a fierce battle, the soldiers wielding swords and spears. The general turns away a machine-gun salesman. He complains: "I don't have time to talk to you—can't you see we're fighting?"

Why are so many firmware projects so late and so bug-ridden? A lot of theories abound about software's complexity and other contributing factors, but I believe the proximate cause is that coding is not something suited to homo sapiens. It requires a level of accuracy that is truly super-human. And most of us ain't Superman.

Cavemen did not have to bag every gazelle they hunted—just enough to keep from starving. Farmers never expect the entire bag of seeds to sprout; a certain wastage is implicit and accepted. Any merchant providing a service expects to delight most, but not all, customers.

The kid who brings home straight As (not mine, darn it) thrills his parents. Yet we get an A for being 90% correct. Perfection isn't required. Most endeavors in life succeed if we score an A, if we miss our mark by 10% or less.

Except in software. 90% correct is an utter disaster, resulting in an unusable product. 99.9% correct means we're shipping junk. 100K lines of code with 99.9% accuracy suggests some 100 lurking errors. That's not good enough. Software requires near-perfection, which defies the nature of intrinsically error-prone people.

Software is also highly entropic. Anyone can write a perfect 100 line-of-code system, but as the size soars perfection, or near perfection, requires ever-increasing investments of energy. It's as if the bits are wandering cattle trying to bust free from the corral; coding cowboys work harder and harder to avoid strays as the size of the herd grows.

So what's the solution? Is there an answer?

In my opinion, software will always be a problem, and there will never be a silver bullet that satisfies all of the stakeholders. But there are some well-known, though rarely practiced, strategies that offer hope.

Some of those are outlined in this chapter. Take some time to study these.

Stop bailing and plug the leaks.

Tools and Methods for Improving Code Quality

By Chris Keydel and Olaf Meding
info@esacademy.com
Embedded Systems Academy
www.esacademy.com

Christian Keydel is a director of the Embedded Systems Academy, where he supervises new class development and consults clients on embedded technologies including CAN and CANopen. He is a frequent speaker at the Embedded Systems Conferences and the Real-Time and Embedded Computing Conferences.

Olaf Meding occasionally teaches as a freelance tutor for Embedded Systems Academy. Olaf has fifteen years of experience in all aspects of software design and development. Previously, he was the principal software developer of a NASA project capable of growing plants in space and he managed a 40-person year mission critical client server application. Olaf has a BSEE from the University of Madison, Wisconsin.

Introduction

These days, there seems to be a shortage of everything. Not enough housing, not enough money, not enough lanes on the freeways—and certainly and above all there's never enough time. This also holds true for the time that is available to develop an embedded system. Two time-consuming tasks are the software and the hardware development. This paper shows some ways not only to reduce the time needed for two very time-consuming tasks—hardware and software design—it also shows how to improve code quality at the same time.

The Traditional Serial Development Cycle of an Embedded Design

Before we can improve the development cycle for an embedded system, we need to analyze the entire development cycle briefly to get a better understanding about how much time is spent—and often wasted—in the different stages of that development cycle while trying to achieve a certain product quality.

One crucial mistake: because everybody knows that time is so valuable these days, the development of a new product is often started too hastily, so that in the end everything takes much longer than it should have.

Very often a new product starts with a great idea. This idea is then quickly (often too quickly) turned into a rough specification that is presented to marketing and/or management. After some meetings and some market research, the decision is made to develop the product. Then the marketing and promotion of the product may start immediately, often still based on the original draft specification. Meanwhile, engineering is working hard to get out that first prototype of the new device. Once available, the prototype is presented to manufacturing—which might look at it and say that the device is not manufacturable like this!

So, more engineering cycles are triggered, trying to make the system ready for volume production while at the same time trying to implement a number of changes requested by different people from different departments (including management). Each cycle takes quite a while, and when the finished product is finally released, chances are it is incapable of what the original specification outlined, plus it is several months late and costs much more to make than anticipated.

And this outline is not fiction. There are multi-billion dollar companies that make all of the mistakes outlined above, spending millions of dollars to develop something they cannot manufacture.

Typical Challenges in Today's Embedded Market

It seems to be a trend in today's marketplace: everything is getting shorter—except maybe the time you are expected to spend in your cubicle, or the list of your competitors.

It is definitely getting harder to meet the customer's demands on time just because there IS no time. The time customers are willing to wait until a new product is available has decreased significantly over the last decade. The reason is simple: technology has advanced so much that product life cycles are much shorter than they used to be, which in turn drastically reduces the window of opportunity that a device needs to hit in order to generate the maximum return (or any return at all) on the investment of its development. Therefore it is only natural that the development time of the products needs to be reduced significantly as well, while still making sure that it is possible to manufacture them in volume production. The better the overall timing, the better the profit margin.

Probably the most drastic example is the hard disk drive industry, where the window of opportunity is a mere 6-8 weeks. If a new hard drive is delayed by 6 weeks, the entire window of opportunity may have passed and it will compete with the next generation of products from other manufacturers. Result: Delays cost real money!

Although this is an extreme example, even areas with traditionally long development cycles like the automotive industry cannot escape the trend of shorter product life cycles.

Generic Methods to Improve Code Quality and Reduce the Time-to-Market

There are a few generic things that can be done to improve the design cycle of new products, which will be discussed in the following paragraphs. Some of the topics mentioned will be discussed in more detail later on in this paper.

Freeze the Specification and Work In Parallel

One of the most important issues is that ALL departments need to be involved in the specification of the new device, including marketing, sales, documentation, manufacturing, and quality assurance, maybe even order processing. This process is called parallel or concurrent development (in contrast to the serial approach described above). This holds true even more if it is a product that can make or break the company depending on its success. After all departments agree on the final specification, it needs to be frozen and must not be changed, even if someone comes up the next day with a new feature that "must be implemented immediately." Collect these sorts of inputs for the next product revision.

Create Milestones

Only with a final, frozen specification, hardware and software development and the work on the documentation can start in parallel. Be sure to set enough milestones for component and module integration. One important difference to the traditional (serial) way of developing a product becomes visible here: it's all about timing, not just speed. Where do we have milestones from different departments that can be merged at certain points in the development cycle?

If you start off too hastily, you might have a prototype faster, but in the long run you might lose valuable time before the product hits the shelves.

Use Available Resources

During the development cycle, do not re-invent the wheel. A good example is not trying to develop your own Real-Time Operating System (RTOS) on the microcontroller you picked. (If no commercial RTOS is available for your microcontroller, you might want to rethink your choice.) Also, do not develop your own networking protocol. There are plenty of well-proven protocols available for different requirements (e.g., CAN (Controller Area Network), I²C (Inter-Integrated Circuit) bus, USB (Universal Serial Bus), RS485, or just the good old RS232). Just make sure your microcontroller has the right interface on-chip.

Don't be cheap about development tools, even if they increase your up-front costs. Engineering time is very expensive as well, and you can cut it down to a fraction if you give your teams the right tools to get the job done faster; some examples being code generator tools or in-circuit emulators.

Provide Continuing Education

Today, technology develops so fast that self-learning or learning-by-doing is inefficient and eats up too much valuable engineering time. Just buying that in-circuit emulator might not be enough if its complex debugging features are not used because no one in the development team knows they exist or how to use them. Expert knowledge on the available resources, especially hardware and software development tools, is essential to get the most benefits from them. A few days of training on the chosen microcontroller platform and the development tools go a long way towards a shorter development time—and more motivated employees. Professional product training is also a great way to get new team members up to speed fast.

Major Time Factors for the Engineering Cycle

The total engineering time needed to implement an embedded application with a frozen specification can be divided into four main sections:

NOTE: We did not include end-customer documentation, as with a good, frozen specification that can be written completely in parallel to the engineering cycle.

The four main sections of the engineering cycle are:

1. Hardware development
2. Software development
3. System integration: Time needed to integrate all components of the application (including potential third party products)
4. Test and quality assurance: Time needed to test the product

A potential fifth section is debugging, however debugging is not really a stage by itself, it is part of all the four sections listed above.

Where is Most Time Needed?

Embedded Applications differ a lot from each other and it is almost impossible to generalize and specify which stages take up how much percent of the total development time. Here's an estimate:

1. Hardware development: 10% to 40%
2. Software development: 20% to 50% or more (potentially grows in the future)
3. System integration: 10% to 40%
4. Test and quality assurance: 10% to 50%
5. Overall debugging 25% to 75%

How to Improve Software Development Time and Code Quality

Before we can actually discuss ways to improve the way we write the software for an application, we need to look at some of the challenges. The programming skills needed to develop an embedded application are getting more and more diverse. The software engineer must be able to master interrupt structures with multiple priority levels and needs to know how to perform high-level software interfacing to a Real-Time Operating System (RTOS), device libraries and networks. By the same token, he/she actually needs some hardware skills to integrate programmable logic and to code the parts of the software interfacing to the underlying hardware components. The next paragraphs will highlight some ways to shorten the software development time.

Write Code in Accordance to an In-house Software Style Guide

Using a software style guide becomes mandatory for companies that seek some sort of quality assurance certification. The idea behind using a software style guide is to standardize the way software is written. It eliminates some of the "creative freedom" programmers have like squeezing in as many instructions as possible into a single line. One of the main benefits is that code becomes easier to read as the reader does not need to get get used to the individual preferences of each programmer.

Unfortunately there is no single guide for the C programming language—there are hundreds, most of them in-house guides, confidential and not publicly available.

Let's look at some rules that can be found in such guides. One of them might be that the length of hexadecimal and binary values directly written into the code must match the data type they are assigned to. For instance, a 16-bit unsigned integer should be assigned a 4 digit hexadecimal number—so 0x000F instead of 0x0F.

The naming rule for procedures and functions could state that the prefix for each name indicates the module this function or procedure is implemented in. The name itself should be a verb followed by an object. So "Com_InitChannel" would be a procedure from module "Com" (communication). If a value is returned, the function name should reflect the return value: "Com_GetErrorCode."

Another section of a style guide usually deals with the comments and could specify when to use comments at the end of a line and when to use a comment line. Typically end of line comments are avoided, as they are more time-consuming to maintain. Blocks with comments should not use any formatting characters (like "*" or "|") at the beginning or at the end of the line as they are also too time consuming to maintain.

When it comes to formatting code blocks using "{" and "}" and indention one of the more often adopted formats is:

```
statement
{
   statements;
}
```

versus

```
statement
   {
   statements;
   }
```

or

```
statement {
   statements;
}
```

Do Code Reviews

Doing a code review means that each line of code written by any programmer needs to be reviewed and understood—in short, confirmed by another programmer. These reviews should begin right after the start of development. While this seems like redundant work at first, it has some very important advantages in the long run. Every bug in the code that is found before starting the real debugging process can save hours of debugging time. So it helps to minimize the hardly predictable time needed for debugging, by adding a little clearly computable review time.

There is also a nice side effect to code reviews: It allows for cross training of the software engineers! If one of the engineers becomes unavailable for some reason, the others can close the gap more easily if they participated in code reviews.

Pick Appropriate Development Tools

An *integrated development environment* with a graphical user interface that ties together software tools like compiler, assembler, linker and locator and maybe also already includes a simulator or a monitor debugger should be a given. But there are several additional development tools that can definitely improve the time it takes to get your application code written faster and to make it more reusable and more reliable at the same time.

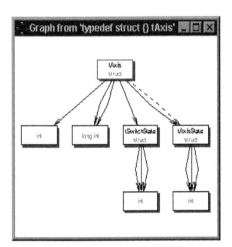

Figure 4-1.

Using a *code editor* provides a faster development environment and ensures greater programmer productivity by adding colors for different keywords, automatic formatting, navigation features and a high level of customization. Adding a *code browser* (Figure 4-1) makes it a lot easier to understand and analyze existing code by allowing browsing code by symbols like functions, variables, constants, types and macros.

One example for such a tool is the "Development Assistant for C" (DA-C) from RistanCASE GmbH in Switzerland. With one click, the programmer is able to view the definition of a variable, all lines in which the variable is used as well as all lines in which the variable is written to. Being able to view calling trees for functions and graphs explaining the composition of complex data structures are extremely helpful especially when making code

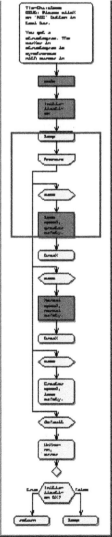

Figure 4-2.

enhancements. This tool also contains a *C syntax analyzer* on a project level—no compiler is needed to make sure your code complies with "Safe C" standards and matches specified software metrics.

The icing on the cake, however, are the *flow charts* (see Figure 4-2) generated by DA-C based on the code and its comments. This feature allows you to browse the source lines and the flow chart in parallel and even to use a third-party debugger to debug your code based on the flow chart. The flow chart to the right was generated by DA-C.

A *version control system* (e.g., MKS's Source Integrity or Microsoft's Visual SourceSafe) can dramatically improve software team coordination—making sure, by default, that only one person at a time is modifying a file. This prevents files from accidentally being replaced by another user's changes. Usually this default can be changed to allow multiple simultaneous checkouts of a single file, while still preventing overwrites of other changes. Old versions of source code and other files can be archived and can be retrieved for bug tracking and other purposes. Version control systems also allow you to track which programs use which modules so that code can be reused more easily. The second half of this document, Part II, will talk about it's features and benefits extensively.

Code generator tools, running locally—like Infineon's Digital Application Engineer DAvE or IAR's MakeApp—or online like ESAcademy's CodeArchitect (www.codearchitect.org) let the user configure a selected microcontroller visually using simple configuration dialog boxes. Optimized source code can then be created automatically to implement device drivers for the configured peripheral modules, saving a large amount of valuable software development time.

Another tool, which requires some financial investment up front, but can pay for itself quite fast, is the *in-circuit emulator* (ICE). In comparison to a standard debugger it offers some unbeatable features. As there is no standard definition of what an ICE is and because there are systems from $300 to $30,000 or more available, we would like to point out a few features you should be looking for, when selecting a system.

- An ICE should be "real real-time" by not stealing any cycles from the microcontroller's CPU.

- No other resources (e.g., on-chip serial channels or timers) should be required for debugging.

- The emulation memory can be mapped
    - Preferably it should be a dual-ported memory allowing access without real-time violation while the code is running.

- On-board trace memory shows the execution history
    - High-end systems include time stamps and/or external signals.

    - High-end systems can filter the race recording, allowing one to record only certain bus accesses or certain code regions (e.g., tasks, functions or interrupt device routines).

- Complex trigger events (like triggering on certain bus states) make debugging faster and easier.

- Memory protection functionality protects selected memory areas from certain accesses (e.g., no code fetch from data or unused addresses).

- A detailed performance analysis on minimum, maximum and average execution times for selected functions can be achieved.

- High-end systems can detect violations in specified cycle times. These systems can break, if an interrupt takes longer to execute than time x or if it does not appear within timeframe y.

Re-using instead of re-inventing

Getting an off-the-shelf *Real-Time Operating System (RTOS)* can save a lot of software engineering hours that you would otherwise have to spend on developing a software structure on your microcontroller platform that allows multiple tasks to be active in parallel. Many RTOSs are available today for a variety of microcontrollers. Many of them don't charge royalty fees and include the entire source code. Just make sure your RTOS is truly preemptive, meaning that it can interrupt a running task A to start another higher priority task B before task A is finished.

If your application has distributed control units you will need a *communications protocol* to exchange information between the units. But do yourself a favor and don't invent your own proprietary protocol when there are so many proven protocols already out there to choose from. Take I²C for example for low-level, on-board communication, or RS232 (UART) for communication between just a few nodes. If you need reliable communication protocol implementing a network of nodes, Controller Area Network (CAN) could be the right choice. Add a higher-layer CAN protocol like CANopen or DeviceNet if you need to ensure data object oriented communication. For applications involving embedded internetworking, TCP/IP should be your protocol of choice.

How to Reduce Hardware Development Time

As mentioned earlier, hardware development time is primarily a factor for embedded systems designs with high volume.

Let's again point out some requirements for your engineering team. There are a variety of different components your team needs to be familiar with. They need to know how to compose the application from ingredients like intelligent devices (microcontroller, microprocessor or DSP), memory (e.g., Flash, EEPROM, RAM), programmable logic, I/O components and analog components. In addition, a networking infrastructure might be needed. Towards the software group an adequate software interface needs to be provided. The different skills required vary so much, that different engineers with different expertise areas might be required.

The number one recommendation to shorten hardware development time is to avoid it:

Use as Many Off-the-Shelf Products as Feasible

Usage of off-the-shelf products in form of single board computers (SBC) drastically shortens the hardware development time. In the past, such boards were only available in the high-performance level, getting close to embedded PCs.

These days, there are several companies like PHYTEC, which focus on the lower to medium range of performance by offering SBCs based on 8-bit or 16-bit microcontrollers. These boards are usually designed on multi-layer boards with all the main circuitries required to get the microcontroller running. Your application might be designed and completed by just adding a 2-layer board with the I/O specific to your application.

Choose Your Microcontroller Wisely

When you decide on the microcontroller architecture for your next application, carefully evaluate the answers to the following questions:

- What development tools are available for this architecture?

- How much experience does the team have using this device—and the development tools?

- How many components / peripherals are on-chip, how many are needed in addition externally? (Each additional component adds cost and time!)

- Are there second sources (other manufacturers) for this chip?

In general, it makes sense to stay with an architecture your team is familiar with for as long as possible. If you must make a transition, ensure that the microcontroller of choice has been around for a while and that the development tools (compilers and in-circuit emulators) are available from several third party vendors.

Example for Microcontrollers that allow shortest time-to-market—Philips

One type of hardware that is frequently overlooked but adds cost and time to the development of the hardware is glue logic. Glue logic requires additional components, occupies valuable PCB space and increases the chances for making design errors due to the additional circuitry. A good example how to get rid of glue logic is the Philips 8051-compatible 51LPC series of microcontrollers. This product family requires no external clock and no external reset circuitry. It offers internal code memory and I/O pins that are strong enough to drive LEDs. All this saves external components—and time.

Only two pins are required to start and operate the 51LPC: Ground and VCC, all others can be used as I/O.

Outlook and Summary

Unfortunately there is no magic when it comes to reducing the time-to-market, as there is no single solution that miraculously cuts your time-to-market in half. Only the combination of many or all of the little things we recommended will help you to shorten the development time, or at a minimum make it more predictable.

If we look into the future with the trend of more and more features (requiring more software) going into embedded systems, then the issue of how to develop better software faster becomes more and more important.

In that regard it is helpful that we currently see an emerging chip architecture that has the potential of becoming a standard for embedded applications. The current huge selection of different architectures from different manufacturers makes it difficult for design teams to select one and stay with it for all projects.

ARM definitely has the potential to change this. As ARM is licensed and manufactured by many different chip manufacturers, it has the potential of becoming a standard controller for embedded applications requiring 32-bit power.

On the low-end, we still see the 8051 as being one of the best choices. With the giant selection of parts from many different suppliers and well-established tools a matching derivative can be found for almost every application that can be handled by an 8-bit microcontroller. And if a supplier fails you, a similar device from another manufacturer might be just around the corner.

Links to companies and or products mentioned in this chapter:

CodeArchitect
On-line code generator tool for microcontrollers
www.codearchitect.org

EmbeddedLinks
Database of links relevant to the Embedded Marketplace and event calendar
www.embeddedlinks.com

Embedded Systems Academy
On-line training classes, multi-day training classes on microcontroller architectures and embedded networking technologies
www.esacademy.com

Infineon Technologies—DavE
Code generator for Infineon Microcontrollers
www.infineon.com/dave/

Philips Semiconductors
8-bit and 16-bit microcontrollers, 51LPC—fastest time-to-market microcontroller
www.philipsmcu.com

Phytec
Single board computers with 8-bit and 16-bit microcontrollers
www.phytec.com

RistanCASE
Code browser: Development Assistant – C
www.ristancase.ch

Triscend
Microcontrollers with on-chip FPGA for customized peripherals
www.triscend.com

Tips to Improve Functions

Jack Ganssle

It's possible to write ugly, obfuscated, horribly convoluted and undocumented code that works. God knows there's an awful lot of it out there today, controlling everything from disposable consumer products to mission-critical avionics. But it's equally possible, and quite a bit easier, to write finely-crafted functions whose inherent correctness shows through, and which are intrinsically maintainable. Here are some guidelines.

Minimize Functionality

Are you an EE? My informal surveys suggest that around 60–70% of all firmware folks have electrical engineering degrees. That background serves us well in understanding both the physics of our applications as well as the intricacies of the hardware we control.

Yet most EE curricula ignore software engineering. Sure, the professors teach you to program, and expect each student to be very proficient at building code. But they provide a didactic devoid of the critical tenets of software engineering necessary to building large, reliable, systems. The skills needed to create a working 500-line program do not scale to one of 100,000 lines.

Probably the most well-known yet least used rule of software design is to keep functions short. In my firmware lectures I ask how many attendees have an *enforced* function size limitation. It's rare to find even a single hand raised. Yet we know that good code is virtually impossible with long routines.

If you write a function that exceeds 50 lines—one page—it's too long. Fact is, you probably cannot remember a string of more than 8 or 10 numeric digits for more than a minute or so; how can you expect to comprehend the thousands of ASCII characters that make up a long function? Worse, trying to follow program flow that extends across page boundaries is tough to impossible, as we flip pages back and forth to understand what a gaggle of nested loops are doing.

Keep functions short. That implies that a function should do just one thing. Convoluted code that struggles to manage many disparate activities is too complex to be reliable or maintainable. I see far too many functions that take 50 arguments selecting a dozen interacting modes. Few of these work well.

Express independent ideas independently, each in its own crystal clear function. One rule of thumb is that if you have trouble determining a meaningful name for a function, it's probably doing too many different things.

Encapsulate

OOP advocates chant the object mantra of "encapsulation, inheritance and polymorphism" like Krishna devotees. OOP is a useful tool that can solve a lot of problems, but it's not the only tool we possess. It's not appropriate for all applications.

But encapsulation is. If you're so ROM-limited that every byte counts, encapsulation may be impossible. But recognize that those applications are inherently very expensive to build. Obviously in a few extremely cost-sensitive applications—like an electronic greeting card—it's critically important to absolutely minimize memory needs. But if you get into a byte-limited situation, figure development costs will skyrocket.

Encapsulation means binding the data and the code that operates on that data into a single homogeneous entity. It means no other bit of code can directly access that data.

Encapsulation is possible in C++ and in Java. It's equally available to C and assembly programmers. Define variables within the functions that use them, and ensure their scope limits their availability to other routines.

Encapsulation is more than data hiding. A properly encapsulated object or function has high cohesion. It accomplishes its mission completely, without doing unrelated activities. It's exception-safe and thread-safe. The object or function is a completely functional black box requiring little or no external support.

A serial handler might require an ISR to transfer received characters into a circular buffer, a "get_data" routine that extracts data from the data structure, and an "is_data_available" function that tests for received characters. It also handles buffer overrun, serial dropout, parity errors, and all other possible error conditions. It's reentrant so other interrupts don't corrupt its data.

A collarary of embracing encapsulation is to delete dependencies. High cohesion must be accompanied by low coupling—little dependence on other activities. We've all read code where some seemingly simple action is intertwined in the code of a dozen other modules. The simplest design change requires chasing variables and functionality throughout thousands of lines of code, a task sure to drive maintainers to drink. I see them on the street here in Baltimore by the dozen, poor souls huddling in inadequate coats, hungry and unshaven, panhandling for applets. If only they'd "said no" to the temptation of global variables.

Remove Redundancies

Get rid of redundant code. Researchers at Stanford studied 1.6 million lines of Linux and found that redundancies, even when harmless, highly correlate with bugs (see www.stanford.edu/~engler/p401-xie.pdf).

They defined redundancies as code snippets that have no effect, like assigning a variable to itself, initializing or setting a variable and then never using that value, dead code, or complex conditionals where a subexpression will never be evaluated, since its logic is already part of a prior subexpression. They were clever enough to eliminate special cases like setting a memory mapped I/O port, since this sort of operation looks redundant but isn't.

Even harmless redundancies that don't create bugs are problems, since these functions are 50% more likely to contain hard errors than other functions that do not have redundant code. Redundancy suggests the developers are confused, and so are likely to make a series of mistakes.

Watch out for block-copied code. I am a great believer in reuse, and encourage the use of previously-tested chunks of source. But all too often, developers copy code without studying all of the implications. Are you really sure all of the variables are initialized as expected, even when this chunk is in a different part of the program? Will a subtle assumption about a mutex create a priority inversion problem?

We copy code to save development time, but remember there is a cost involved. Study that code even more carefully than the new stuff you're writing from scratch. And when Lint or the compiler warns about unused variables, take heed: this may be a signal there are other, more significant, errors lurking.

Reduce Real-Time Code

Real-time code is error-prone, expensive to write, and even more costly to debug. If it's at all possible, move the time-critical sections to a separate task or section of the program. When time issues infiltrate the entire program, then every bit of the program will be hard to debug.

Today we're building bigger and more complex systems than ever, with debugging tools whose capabilities are less than those we had a decade ago. Before processor speeds zoomed to near infinity and the proliferation of SMT packages eliminated the ability of coffee drinkers to probe ICs, the debugger of choice was an in-circuit emulator. These included real-time trace circuits, event timers, and even performance analyzers. Today we're saddled with BDM or JTAG debuggers. Though nice for working on procedural problems, they offer essentially no resources for dealing with problems in the time domain.

Remember also two rules of thumb: a system loaded to 90% doubles development time over one at 70% or less. At 95% the schedule triples. Real-time development projects are expensive; highly loaded ones even more so.

Flow With Grace

Flow, don't jump. Avoid continues, gotos, breaks and early returns. These are all useful constructs, but generally reduce a function's clarity. Overused, they are the basic fabric of spaghetti code.

Refactor Relentlessly

XP and other agile methods emphasize the importance of refactoring, or rewriting crummy code. This is not really a new concept; Capers Jones, Barry Boehm and others have shown that badly-written modules are much more expensive to beat into submission and maintain than ones with a beautiful structure.

Refactoring zealots demand we rewrite any code that can be improved. That's going too far, in my opinion. Our job is to create a viable product in a profitable way; perfection can never be a goal that overrides all other considerations. Yet some functions are so awful they must be rewritten.

If you're afraid to edit a function, if it breaks every time you modify a comment, then it needs to be refactored. Your finely-honed sense as a professional developer that, well, we just better leave this particular chunk of code intact because no one dares mess with it, is a signal that it's time to drop everything else and rewrite the code so it's understandable and maintainable.

The second law of thermodynamics tells us that any closed system will head to more disorder; that is, its entropy will increase. A program obeys this depressing truth. Successive maintenance cycles always increase the software's fragility, making each additional change that much more difficult. As Ron Jeffries pointed out, maintenance without refactoring increases the code's entropy by adding a "mess" factor (m) to each release. The cost to produce each release looks something like: $(1+m)(1+m)(1+m)....$, or $(1+m)^n$, where n is the number of releases. Maintenance costs grow exponentially as we grapple with more and more hacks and sloppy shortcuts. This explains that bit of programmer wisdom that infuriates management: "the program is too much of a mess to maintain."

Refactoring incurs its own cost, r. But it eliminates the mess factor, so releases cost $1+r+r+r...$, which is linear.

Luke Hohmann advocates "post release entropy reduction." He recognizes that all too often we make some quick hacks to get the product out the door. These entail a maintenance cost, so it's critical we pay off the technical debt incurred in being abusive to the software. Maintenance is more than cramming in new features; it's also reducing accrued entropy.

Refactor to sharpen fuzzy logic. If the code is a convoluted mess or just not absolutely clear, rewrite it so it better demonstrates its meaning. Eliminate deeply nested loops or conditionals—no one is smart enough to understand all permutations of IFs nested 5 levels deep. Clarity leads to accuracy.

Employ Standards and Inspections

Write code using your company's firmware standard. Use formal code inspections to ensure adherence to the standard and to find bugs. Test only after conducting an inspection.

Inspections are some 20 times cheaper at finding bugs than traditional debugging. They'll capture entire classes of problems you'll never pick up with conventional testing. Most studies show that traditional debugging checks only about half the code! Without inspections you're most likely shipping a bug-ridden product. It's interesting that the DO-178B standards for building safety critical software rely heavily on the use of tools to insure every line of code gets executed. These code coverage tools are a wonder, but are no substitute for inspections.

Without standards and inspections it's just not possible to habitually build great firmware products.

Comment Carefully

Even if you're stuck in a hermetically sealed cubicle, never interacting with people and just cranking code all day, I contend you still have a responsibility to communicate clearly and grammatically with others. Software is, after all, a mix of computerese (the C or C++ itself) and comments (in America, at least, an English-language description meant for humans, not the computer). If we write perfect C with illegible comments, we're doing a lousy job.

I read a *lot* of code from a huge range of developers. Consistently well-done comments are rare. Sometimes I can see the enthusiasm of the team at the project's outset. The startup code is fantastic. Main()'s flow is clear and well documented. As the project wears on functions get added and coded with less and less care. Comments like

```
/* ???? */
```

or my favorite:

```
/* Is this right? */
```

start to show up. Commenting frequency declines; clarity gives way to short cryptic notes; capitalization descends into chaotic randomness. The initial project excitement, as shown in the careful crafting of early descriptive comments, yields to schedule panic as the developers all but abandon anything that's not executable.

Onerous and capricious schedules are a fact of life in this business. It's natural to chuck everything not immediately needed to make the product work. Few bosses grade on quality of the source code. Quality, when considered at all, is usually a back-end complaint about all the bugs that keep surfacing in the released product, or the ongoing discovery of defects that pushes the schedule back further and further.

My standard for commenting is that someone versed in the functionality of the product—but not the software—should be able to follow the program flow by reading the comments without reference to the code itself. Code implements an algorithm; the comments communicate the code's operation to yourself and others. Maybe even to a future version of yourself during maintenance years from now.

Write every bit of the documentation (in the USA at least) in English. Noun, verb. Use active voice. Be concise; don't write the Great American Novel. Be explicit and complete; assume your reader has not the slightest insight into the solution of the problem. In most cases I prefer to incorporate an algorithm description in a function's header, even for well-known approaches like Newton's Method. A description that uses your variable names makes a lot more sense than "see any calculus book for a description." And let's face it: once carefully thought out in the comments it's almost trivial to implement the code.

Capitalize per standard English procedures. IT HASN'T MADE SENSE TO WRITE ENTIRELY IN UPPER CASE SINCE THE TELETYPE DISAPPEARED 25 YEARS AGO. the common practice of never using capital letters is also obsolete. Worst aRe the DevElopeRs wHo usE rAndOm caSe changeS. Sounds silly, perhaps, but I see a lot of this. And spel al of the wrds gud.

Avoid long paragraphs. Use simple sentences. "Start_motor actuates the induction relay after a three second pause" beats "this function, when called, will start it all off and flip on the external controller but not until a time defined in HEADER.H goes by."

Begin every module and function with a header in a standard format. The format may vary a lot between organizations, but should be consistent within a team. Every module (source file) must start off with a general description of what's in the file, the company name, a copyright message if appropriate, and dates. Start every function with a header that describes what the routine does and how, goes-intas and goes-outas (i.e., parameters), the author's name, date, version, a record of changes with dates and the name of the programmer who made the change.

C lends itself to the use of asterisks to delimit comments, which is fine. I see a lot of this:

```
/ * * * * * * * * * * * * *

*    comment       *

* * * * * * * * * * * * * /
```

which is a lousy practice. If your comments end with an asterisk as shown, every edit requires fixing the position of the trailing asterisk. Leave it off, as follows:

```
/ * * * * * * * * * * * * *

*    comment

* * * * * * * * * * * * * /
```

Most modern C compilers accept C++'s double slash comment delimiters, which is more convenient than the /* */ C requires. Start each comment line with the double slash so the difference between comments and code is crystal clear.

Some folks rely on a fancy editor to clean up comment formatting or add trailing asterisks. Don't. Editors are like religion. Everyone has their own preference, each of which is configured differently. Someday compilers will accept source files created with a word processor which will let us define editing styles for different parts of the program. Till then dumb ASCII text formatted with spaces (not tabs) is all we can count on to be portable and reliable.

Enter comments in C at block resolution and when necessary to clarify a line. Don't feel compelled to comment each line. It is much more natural to comment groups of lines which work together to perform a macro function.

Explain the meaning and function of every variable declaration. Long variable names are merely an *aid* to understanding; accompany the descriptive name with a deep, meaningful, prose description.

One of the perils of good comments—which is frequently used as an excuse for sloppy work—is that over time the comments no longer reflect the truth of the code. Comment drift is intolerable. Pride in Workmanship means we change the docs as we change the code. The two things happen in parallel. Never defer fixing comments till later, as it just won't happen. Better: edit the descriptions first, then fix the code.

One side effect of our industry's inglorious 50-year history of comment drift is that people no longer trust comments. Such lack of confidence leads to even sloppier work. It's hard to thwart this descent into commenting chaos. Wise developers edit the header to reflect the update for each patch, but even better add a note that says "comments updated, too" to build trust in the docs.

Finally, consider changing the way you write a function. I have learned to write all of the comments first, including the header and those buried in the code. Then it's simple, even trivial, to fill in the C or C++. Any idiot can write software following a decent design; inventing the design, reflected in well-written comments, is the really creative part of our jobs.

Summary

When I was just starting my career an older fellow told me what he called The Fundamental Rule of Engineering: if the damn thing works at all, leave it alone. It's an appealing concept, one I used too many times over the years. The rule seems to work with hardware design, but is a disaster when applied to firmware. I believe that part of the software crisis stems from a belief that "pretty much works" is a measure of success. Professionals, though, understand that developmental requirements are just as important as functional and operational requirements. Make it work, make it beautiful, and make it maintainable.

Evolutionary Development

(How to deliver Quality On Time in Software
Development and Systems Engineering Projects)

Niels Malotaux

Niels Malotaux is an independent consultant teaching immediately applicable methods for delivering Quality On Time to R&D and software organizations. Quality On Time is short for delivering the right results, within the time and budget agreed, with no excuses, in a pleasant way for all involved, including the developers. Niels does not just tell stories, he actually puts development teams on the Quality On Time track and coaches them to stay there and deliver their quality software or systems on time, without over-time, without the need for excuses.

Introduction

Software developers systematically fail to manage projects within the constraints of cost, schedule, functionality and quality. More than half of ICT are still not content with the performance of ICT suppliers [Ernst&Young, 2001]. This has been known for some 35 years. Solutions have been developed, with impressive results published years ago (e.g. Mills, 1971 [1], Brooks, 1987 [2], Gilb, 1988 [3]). Still, in practice not much has changed. An important step in solving this problem is to accept that *if developers failed to improve their habits*, in spite of the methods presented in the past, *there apparently are psychological barriers in humans, preventing adoption of these methods*. The challenge is to find ways to catch the *practical essence of the solutions* to manage projects within the constraints of cost, schedule, functionality and quality and *ways* to get the developers to use these solutions.

The importance of solving the problem is mainly economical:

■ Systematically delivering software development results within the constraints of cost, schedule, functionality and quality saves unproductive work, both by the developers and the users (note Crosby, 1996: the Price Of Non-Conformance [4]).

■ Prevention of unproductive work eases the shortage of IT personnel.

■ Enhancing the quality level of software developments yields a competitive edge.

■ Being successful eases the stress on IT personnel, with positive health effects as well as positive productivity effects.

In this chapter, we show methods and techniques, labelled "Evo" (from Evolutionary), which enable software developers and management to deliver "Quality On Time", which is short for successfully managing projects within the constraints of cost, schedule, functionality and quality. These methods are taught and coached in actual development projects with remarkable results.

The contents is based on practical experiences and on software process improvement research and development and especially influenced by Tom Gilb (1988 [3], later manuscripts [5] and discussions).

1. History

Most descriptions of development processes are based on the Waterfall model, where all stages of development follow each other (Figure 6-1). Requirements must be fixed at the start and at the end we get a Big Bang delivery. In practice, hardly anybody really follows this model, although in reporting to management, practice is bent into this model. Management usually expects this simple model, and most development procedures describe it as mandatory. This causes a lot of mis-communication and wastes a lot of energy.

Early descriptions of Evolutionary delivery, then called Incremental delivery, are described by Harlan Mills in 1971 [1] and F.P. Brooks in his famous "No silver bullet" article in 1987 [2]. Evolutionary delivery is also used in Cleanroom Software Engineering [6]. A practical elaboration of Evolutionary development theory is written by Tom Gilb in his book "Principles of Software Engineering Management" in 1988 [3] and in newer manuscripts on Tom Gilb's web-site [16].

Incremental delivery is also part of eXtreme Programming (XP) [15, 17], however, if people claim to follow XP, we hardly see the Evo element practiced as described here.

We prefer using the expression Evolutionary delivery, or Evo, as proposed by Tom Gilb, because not all Incremental delivery is Evolutionary. Incremental delivery methods use cycles, where in each cycle, part of the design and implementation is done. In practice this still leads to Big Bang delivery, with a lot of debugging at the end. We would like to reserve the term Evolutionary for a special kind of Incremental delivery, where we address issues like:

■ Solving the requirements paradox.

■ Rapid feedback of estimation and results impacts.

■ Most important issues first.

■ Highest risks first.

■ Most educational or supporting issues for the development first.

■ Synchronizing with other developments (e.g. hardware development).

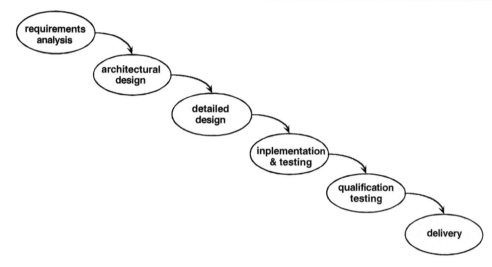

Figure 6-1: Waterfall development model.

- Dedicated experiments for requirements clarification, before elaboration is done.

- Every cycle delivers a useful, completed, working, functional product.

- At the fatal end day of a project we should rather have 80% of the (most important) features 100% done, than 100% of all features 80% done. In the first case, the customer has choice to put the product on the market or to add some more bells and whistles. In the latter case, the customer has no choice but to wait and grumble.

In Evolutionary delivery, we follow the waterfall model (Figure 6-1) repeatedly in very short cycles (Figure 6-2).

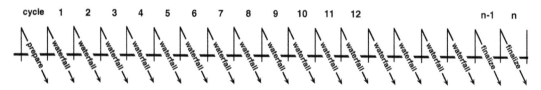

Figure 6-2: Evolutionary delivery uses many waterfalls.

2. Issues Addressed by Evo

A. Requirements Paradoxes

The *1ˢᵗ Requirements Paradox* is:

- Requirements must be stable for reliable results.

- However, the requirements always change.

Even if you did your utmost to get complete and stable requirements, they will change. Not only because your customers change their mind when they see emerging results from the developments. Also the developers themselves will get new insights, new ideas about what the requirements should really be. So, requirements change is a *known risk*. Better than ignoring the requirements paradox, use a development process that is designed to cope with it: Evolutionary delivery.

Evo uses rapid and frequent feedback by stakeholder response to verify and adjust the requirements to what the stakeholders really need most. Between cycles there is a short time slot where stakeholders input is allowed *and requested* to reprioritize the list.

This is due to the *2ⁿᵈ Requirements Paradox*:

- We don't want requirements to change.

- However, because requirements change now is a *known risk*, we try to *provoke* requirements change as early as possible.

We solve the requirements paradoxes by creating stable requirements *during* a development cycle, while explicitly reconsidering the requirements *between* cycles.

B. Very Short Cycles

Actually, few people take planned dates seriously. As long as the end date of a project is far in the future (Figure 6-3), we don't feel any pressure and work leisurely, discuss interesting things, meet, drink coffee, ... (How many days before your last exam did you really start working...?). So at the start of the project we work relatively slowly. When the pressure of the finish date becomes tangible, we start working harder, stressing a bit, making errors causing

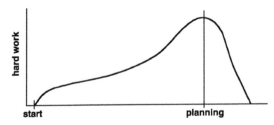

Figure 6-3: We only start working harder when the pressure of the delivery date is near. Usually we are late.

delays, causing even more stress. The result: we do not finish in time. We know all the excuses, which caused us to be late. It's never our own fault. This is not wrong or right. It's human psychology. That is how we function. So don't ignore it. Accept it and then think what to do with it.

Smart project managers tell their team an earlier date (Figure 6-4). If they do this cleverly, the result may be just in time for the real date. The problem is that they can do this only once or twice. The team members soon will discover that the end date was not really hard and they will lose faith in milestone dates. This is even worse.

The solution for coping with these facts of human psychology is to plan in very short increments (Figure 6-5). The duration of these increments must be such that:

- The pressure of the end date is felt on the first day.

- The duration of a cycle must be sufficient to finish real tasks.

Three weeks is too long for the pressure and one week may be felt as too short for finishing real tasks. Note that the pressure in this scheme is much healthier than the real stress and failure at the end of a Big Bang (delivery at once at the end) project. The experience in an actual project, where we got only six weeks to finish completely, led to using one-week cycles. The results were such that we will continue using one-week cycles on all subsequent projects. If you cannot even plan a one-week period, how could you plan longer periods ...?

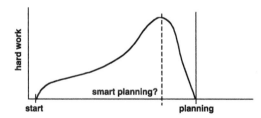

Figure 6-4: To overcome the late delivery problem, a smart project manager sells his team an earlier delivery date. Even smarter developers soon will know.

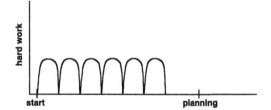

Figure 6-5: The solution: choose short, realistic "delivery dates." Satisfaction, motivation, fast feedback.

C. Rapid and Frequent Feedback

If everything was completely clear we could use the waterfall development model. We call this *production* rather than development. At the start of a new development, however, there are many uncertainties we have to explore and to change into certainties. Because even the simplest development project is too complex for a human mind to oversee completely (E. Dijkstra, 1965: "The competent programmer is fully aware of the limited size of his own skull" [12]) we must learn iteratively what we are actually dealing with and learn how to perform better.

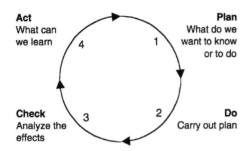

Figure 6-6: Shewhart cycle,
Deming cycle, PDCA cycle.

This is done by "think first, then do," because thinking costs less than doing. But, because we cannot foresee everything and we have to assume a lot, we constantly have to check whether our thoughts and assumptions were correct. This is called feedback: we plan something, we do it as well as we can, then we check whether the effects are correct. Depending on this analysis, we may change our ways and assumptions. Shewhart already described this in 1939 [13]. Deming [14] called it the Shewhart cycle (Figure 6-6). Others call it the Deming cycle or PDCA (Plan-Do-Check-Act) cycle.

In practice we see that if developers do something (section 2 of the cycle), they sometimes plan (section 1), but hardly ever explicitly go through the analysis and learn sections. In Evo we do use all the sections of the cycle deliberately in rapid and frequent feedback loops (Figure 6-7):

- The weekly task cycle

  In this cycle we optimize our estimation, planning and tracking abilities in order to better predict the future. We check constantly whether we are *doing* the right things in the right order to the right level of detail for the moment.

- The frequent stakeholder value delivery cycle

  In this cycle we optimize the requirements and check our assumptions. We check constantly whether we are *delivering* the right things in the right order to the right level of detail for the moment. Delivery cycles may take 1 to 3 weekly cycles.

- The strategic objectives cycle

  In this cycle we review our strategic objectives and check whether what we do still complies with the objectives. This cycle may take 1 to 3 months.

- The organization roadmap cycle

  In this cycle we review our roadmap and check whether our strategic objectives still comply with what we should do in this world. This cycle may take 3 to 6 months.

Figure 6-7:
Cycles in Evo.

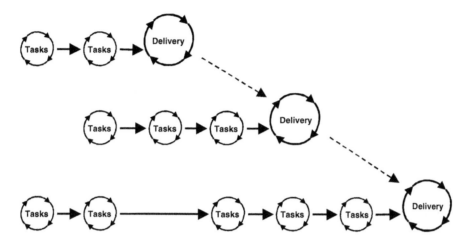

*Figure 6-8: Current tasks feed the current delivery cycle,
as well as prepare for future delivery cycles.*

In development *projects*, only task cycles and delivery cycles are considered. In a task cycle, tasks are done to feed the current delivery, while some other tasks may be done to make future deliveries possible (Figure 6-8).

D. Time Boxing

Evolutionary project organization uses *time boxing* rather than *feature boxing*. If we assume that the amount of resources for a given project is fixed, or at least limited, it is possible to realize either:

- A fixed set of features in the time needed to realize these features. We call this *feature boxing*.

- The amount of features we can realize in a fixed amount of time. We call this *time boxing*.

To realize a fixed set of features in a fixed amount of time with a given set of resources is only possible if the time is sufficient to realize all these features. In practice, however, the time allowed is usually insufficient to realize all the features asked: *What the customer wants, he cannot afford.* If this is the case, we are only fooling ourselves trying to accomplish the impossible (Figure 6-9). This has nothing to do with lazy or unwilling developers: if the time (or the budget) is insufficient to realize all the required features, they *will not all be realized*. It is as simple as that.

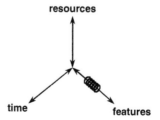

Figure 6-9: If resources and time are fixed, the features are variable.

The Evo method makes sure that the customer gets the most and most important features *possible* within a certain amount of time and with the available resources. Asking developers to accomplish the impossible is one of the main energy drains in projects. By wasting energy the result is always less than otherwise possible.

In practice, time boxing means:

- A set number of hours is reserved for a task.
- At the end of the time box, the task should be 100% done. That means really *done*.
- Time slip is not allowed in a time box, otherwise other tasks will be delayed and this would lead to uncontrolled delays in the development.
- Before the end of the time box we check how far we can finish the task. If we foresee that we cannot finish a task, we should define what we know now, try to define what we still have to investigate, define tasks and estimate the time still needed. Preferably, however, we should try whether we could *go into less detail* this moment, actually finishing the task to a *sufficient* level of detail within the time box. A TaskSheet (details see [8]) is used to define:
  - The goal of the task.
  - The strategy to perform the task.
  - How the result will be verified.
  - How we know for sure that the task is really done (i.e. there is really nothing we have to do any more for this task, we can forget about it).

E. Estimation, Planning and Tracking

Estimation, planning and tracking are an *inseparable trinity. If you don't do one of them, you don't* need the other two.

- If you don't estimate, you cannot plan and there is nothing to track.
- If you do not plan, estimation and tracking is useless.
- If you do not track, why should you estimate or plan?

So:

- Derive small tasks from the requirements, the architecture and the overall design.
- Estimate the time needed for every small task.
- Derive the total time needed from:
  - The time needed for all the tasks.
  - The available resources.
  - Corrected for the real amount of time available per resource (nobody works a full 100% of his presence on the project. The statistical average is about 55%. This is one of the key reasons for late projects! [9])

- Plan the next cycle exactly.

- Be sure that the work of every cycle can be done. That means *really* done. Get commitment from those who are to do the real work.

- Plan the following cycles roughly (the planning may change anyway!).

- Track successes and failures. Learn from it. Refine estimation and planning continuously. Warn stakeholders *well in advance* if the target delivery time is changing because of *any* reason.

- There may be various target delivery times, depending on various feature sets.

If times and dates are not important to you (or to management), then don't estimate, plan, nor track: you don't need it. However, if timing is important, *insist* on estimation, planning and tracking. And it is not even difficult, once you get the hang of it.

If your customer (or your boss) doesn't like to hear that you cannot exactly predict which features will be in at the fatal end day, while you *know* that not all features will be in (at a fixed budget and fixed resources), you can give him two options:

- Either to tell him the day before the fatal day that you did not succeed in implementing all the functions.

- Or tell him now (because you already know), and let him every week decide with you which features are the most important.

It will take some persuasion, but you will see that within two weeks you will work together to get the best possible result. There is one promise you can make: The process used is the most efficient process available. In any other way he will never get more, probably less. So let's work together to make the best of it. Or decide at the beginning to add more resources. Adding resources later evokes Brooks Law [9]: "Adding people to a late project makes it later." Let's stop following ostrich-policy, face reality and deal with it in a realistic and constructive way.

F. Difference Between Effort and Lead-time

If we ask software developers to estimate a given task in days, they usually come up with estimates of lead-time. If we ask them to estimate a task in hours, they come up with estimates in effort. Project managers know that developers are optimistic and have their private multiplier (like 2, $\sqrt{2}$, e or π) to adjust the estimates given. Because these figures then have to be entered in project-planning tools, like MS Project, they enter the adjusted figures as lead-time.

The problem with lead-time figures is that these are a mix of two different time components:

- Effort, the time needed to do the work.

- Lead-time, the time until the work is done. Or rather Lead-time minus Effort, being the time needed for other things than the work to be done. Examples of "other things" are: drinking coffee, meetings, going to the lavatory, discussions, helping colleagues, telephone calls, e-mail, dreaming, etc. In practice we use the Effort/Lead-time ratio, which is usually in the range of 50-70% for full-time team members.

Because the parameters causing variation in these two components are different, they have to be kept apart and treated differently. If we keep planning only in lead-time, we will never be able to learn from the tracking of our planned, estimated figures. Thus we will never learn to predict development time. If these elements are kept separately, people can learn very quickly to adjust their effort estimating intuition. In recent projects we found: first week: 40% of the committed work done, second week: 80% done, from the third week on: 100% or more done. Now we can start predicting!

Separately, people can learn time management to control their Effort/Lead-time ratio. Brooks indicated this already in 1975 [9]: *Programming projects took about twice the expected time. Research showed that half of the time was used for activities other than the project.*

In actual projects, we currently use the rule that people select 2/3 of a cycle (26 hours of 39) for project tasks, and keep 1/3 for other activities. Some managers complain that if we give about 3 days of work and 5 days to do the work, people tend to "Fill the time available." This is called Parkinson's Law [10]: "Work expands so as to fill the time available for its completion." Management uses the same reasoning, giving them 6 days of work and 5 days to do it, hoping to enhance productivity. Because 6 days of effort *cannot* be done in 5 days and people have to do, and *will* do, the other things anyway, people will always *fail to succeed* in accomplishing the impossible. What is worse: this causes a constant sense of failure, causing frustration and demotivation. If we give them the amount of work they can accomplish, they will succeed. This creates a sensation of accomplishment and success, which is very motivating. The observed result is that giving them 3 days work for 5 days is *more productive* that giving them 6 days of work for 5 days.

G. Commitment

In most projects, when we ask people whether a task is done, they answer: "Yes." If we then ask, "Is it really done?", they answer: "Well, almost." Here we get the effect that if 90% is done, they start working on the other 90%. This is an important cause of delays. Therefore, it is imperative that we define when a task is really 100% done and that we insist that any task be 100% done. Not 100% is *not* done.

In Evo cycles, we ask for tasks to be 100% done. *No need to think about it any more.* Upon estimating and planning the tasks, effort hours have been estimated. Weekly, the priorities are defined. So, every week, when the project manager proposes any team member the tasks for the next cycle, he should never say "Do this and do that." He should always propose: "Do you still agree that these tasks are highest priority, do you still agree that you should do it, and do you still agree with the estimations?" If the developer hesitates on any of these questions, the project manager should ask why, and *help the developer* to *re-adjust* such that he can give a *full commitment* that he will accomplish the tasks.

The project manager may help the developer with suggestions ("Last cycle you did not succeed, so maybe you were too optimistic?"). He may *never* take over the responsibility for the decision on which tasks the developer accepts to deliver. This is the only way to get true developer commitment. At the end of the cycle the project manager only has to use the mirror. In the mirror the developer can see himself if he failed in fulfilling his commitments. If the project manager decided what had to be done, the developer sees right through the mirror and only sees the project manager.

It is essential that the project manager coaches the developers in getting their commitments right. Use the sentence: "Promise me to do nothing, as long as *that* is 100% done!" to convey the importance of *completely done*. Only when working with real commitments, developers can learn to optimize their estimations and deliver accordingly. Otherwise, they will *never* learn. Project managers being afraid that the developers will do less than needed and therefore giving the developers more work than they can commit to, will never get what they hope for because without real commitment, people tend to do less.

H. Risks

If there are no risks whatsoever, use the waterfall model for your development. If there are risks, which is the case in any new development, we have to constantly assess how we are going to control these risks. Development is for an important part risk-reduction. If the development is done, all risks should have been resolved. If a risk turns out for worse at the end of a development, we have no time to resolve it any more. If we identify the risks earlier, we may have time to decide what to do if the risk turns out for worse. Because we develop in very short increments of one week the risk that an assumption or idea consumes a lot of development time before we become aware that the result cannot be used is limited to one week. Every week the requirements are redefined, based upon what we learnt before.

Risks are not limited to assumptions about the product requirements, where we should ask ourselves:

- Are we developing the right things right?
- When are things right?

Many risks are also about timing and synchronization:

- Can we estimate sufficiently accurately?
- Which tasks are we forgetting?
- Do we get the deliveries from others (hardware, software, stakeholder responses, ...) in time?

Actually the main questions we are asking ourselves systematically in Evo are: *What* should we do, in *which order*, to *which level of detail* for *now*. Too much detail too early means usually that the detail work has to be done over and over again. Maybe the detail work was not done wrong. It only later turns out that it should have been done differently.

I. Team meetings

Conventional team meetings usually start with a round of excuses, where everybody tells why he did not succeed in what he was supposed to do. There is a lot of discussion about the work that was supposed to be done, and when the time of the meeting is gone, new tasks are hardly discussed. This is not a big problem, because most participants have to continue their unfinished work anyway. The project manager notes the new target dates of the delayed activities and people continue their work. After the meeting the project manager may calculate how much reserve ("slack time") is left, or how much the project is delayed if all reserve has already been used. In many projects we see that project-planning sheets (MS Project) are mainly used as wallpaper. They are hardly updated and the actual work and the plan-on-the-wall diverge more and more every week.

In the weekly Evo team meeting, we only discuss new work, never past work. We do not waste time for excuses. What is past we cannot change. What we still should do is constantly re-prioritized, so we always work on what is best from this moment. We don't discuss past tasks because they are finished. If discussion starts about the new tasks, we can use the results in our coming work. That can be useful. Still, if the discussion is between only a few participants, it should be postponed till after the meeting, not to waste the others' time.

J. Magic Words

There are several "magic words" that can be used in Evo practice. They can help us in doing the *right things* in the *right order* to the *right level of detail for this moment.*

- *Focus*
  Developers tend to be easily distracted by many important or interesting things. Some things may even really be important, however, not at this moment. Keeping focus at the current priority goals, avoiding distractions, is not easy, but saves time.

- *Priority*
  Defining priorities and only working on the highest priorities guides us to doing the most important things first.

- *Synchronize*
  Every project interfaces with the world outside the project. Active synchronization is needed to make sure that planned dates can be kept.

- *Why*
  This word forces us to define the reason why we should do something, allowing us to check whether it is the right thing to do. It helps in keeping focus.

- *Dates are sacred*
  In most projects, dates are fluid. Sacred dates means that if you agree on a date, you stick to your word. Or disclose well in advance that you cannot keep your word. With Evo you will know well in advance.

- *Done*
  To make estimation, planning and tracking possible, we must finish tasks completely.

Not 100% finished is *not* done. This is to overcome the "If 90% is done we continue with the other 90%" syndrome.

- **Bug, debug**
  A bug is a small creature, autonomously creeping into your product, causing trouble, and you cannot do anything about it. Wrong. People make mistakes and thus cause defects. The words *bug* and *debug* are dirty words and should be erased from our dictionary. By actively learning from our mistakes, we can learn to avoid many of them. In Evo, we actively catch our mistakes as early as possible and act upon them. Therefore, the impact of the defects caused by our mistakes is minimized and spread through the entire project. This leaves a bare minimum of defects at the end of the project. Evo projects do not need a special "debugging phase."

- **Discipline**
  With discipline we don't mean imposed discipline, but rather what you, yourself, know what is best to do. If nobody watches us, it is quite human to cut corners, or to do something else, even if we know this is wrong. We see ourselves doing a less optimal thing and we are unable to discipline ourselves. If somebody watches over our shoulder, keeping discipline is easier. So, discipline is difficult, but we can help each other. Evo helps keeping discipline. Why do we want this? Because we enjoy being successful, doing the right things.

3. How Do We Use Evo in Projects

In our experience, many projects have a mysterious start. Usually when asked to introduce Evo in a project, one or more people have been studying the project already for some weeks or even months. So in most cases, there are some requirements and some idea about the architecture. People acquainted with planning usually already have some idea about what has to be done and have made a conventional planning, based on which the project was proposed and commissioned.

A. Evo Day

To change a project into an Evo project, we organize an "Evo day", typically with the Project Manager, the architect, a tester and *all* other people of the development team. Stakeholder attendance can be useful, but is not absolutely necessary at the first Evo day, where we just teach the team how to change their ways. During the Evo day (and during all subsequent meetings) a notebook and a LCD projector are used, so that all participants can follow what we are typing and talking about. It is preferable to organize the Evo day outside the company.

The schedule is normally:

Morning

- Presentation of Evo methods [11]: why and how.
- Presentation of the product by the systems architect (people present usually have different views, or even no view, of the product to be developed).

Afternoon

In the afternoon we work towards defining which activities should be worked on in the coming week/cycle. Therefore we do exercises in:

- Defining sub-tasks of at most 26 hours.
  In practice, only a few activities will be detailed. People get tired of this within 20 minutes, but they did the exercise and anyway we don't have time to do it all in one day.

- Estimating the effort of the sub-tasks, in effort-hours, never in days, see "Difference between effort and lead-time" above.

- Defining priorities.

- Listing the tasks in order of priority.

- Dividing top-priority activities, which have not yet been divided into sub-tasks.

- Estimating effort on top-priority sub-tasks if not yet done.

- The team decides who should do what from the top of the list.

- Every individual developer decides which tasks he will be able to deliver done, really done at the end of the cycle. If a commitment cannot be given, take fewer tasks, until full commitment can be given.

At the end of the day everyone has a list of tasks for the coming week, and a commitment that these tasks will be finished completely, while we are sure that the tasks we start working on have the highest priority.

B. Last Day of the Cycle

The last day of a cycle is special and divided into 3 parts (Figure 6-10, next page):

- The project manager visits every developer individually and discusses the results of the tasks. If the commitments could not be met, they discuss the causes: Was the effort estimation incorrect or was there a time-management problem? The developer should learn from the results to do better the next time. After having visited all developers, the project manager has an overview of the status of the project.

- The status of the project is discussed with the customer, product manager, or which-ever relevant stakeholders. Here the Requirements Paradox is handled: during the week, the requirements were fixed, now is the 1 to 2 hours timeslot that the stake-holders may re-arrange the requirements and priorities. At the end of this meeting, the requirements and priorities are fixed again.

- Finally, the project manager defines task-proposals for the developers and discusses these proposals with them individually. Developers agree that these tasks have the highest priority and commit to finishing these tasks during the cycle.

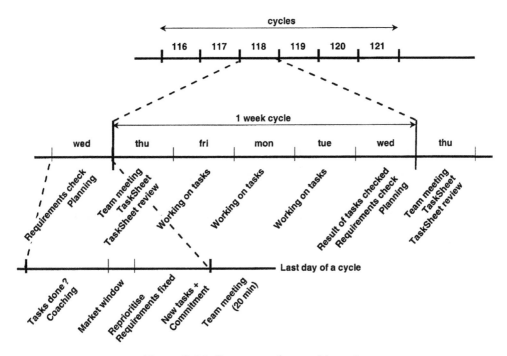

Figure 6-10: Structure of a weekly cycle.

C. Team Meeting

Having prepared the individual task-lists for the next cycle, in the team meeting, at the end of the last cycle day, or the beginning of the first new cycle day, the following is done:

- Experience from the past cycle may be discussed if it could benefit subsequent work.

- The status of the project is discussed. Sub-tasks may be (re-)defined and (re-)estimated if full participation is useful.

- The tasks for the next cycle are formally assigned and committed to. Now all participants hear who is going to do what and may react upon it.

- Discussion may be allowed, if it affects most participants.

- The discussions may cause some reprioritization and thus reshuffling of tasks to be done.

Weekly team meetings typically take less than 20 minutes. A typical reaction at the end of the first Evo team meeting is: "We never before had such a short meeting." When asked "Did we forget to discuss anything important?", the response is: "No, this was a good and efficient meeting." This is one of the ways we are saving time.

4. Check Lists

There are several checklists being used to help define priorities and to help to get tasks really finished. These are currently:

    A. Task prioritization criteria

    B. Delivery prioritization criteria

    C. Task conclusion criteria

A. Task Prioritization Criteria

To help in the prioritization process of which tasks should be done first, we use the following checklist:

- Most important issues first (based on current and future delivery schedules).

- Highest risks first (better early than late).

- Most educational or supporting activities first.

- Synchronization with the world outside the team (e.g. hardware needs test-software, software needs hardware for test: will it be there when needed?).

- Every task has a useful, completed, working, functional result.

B. Delivery Prioritization Criteria

To help in the prioritization process of what should be in the next delivery to stakeholders we use the following checklist:

- Every delivery should have the juiciest, most important stakeholder values that can be made in the least time. Impact Estimation [7] is a technique that can be used to decide on what to work on first.

- A delivery must have *symmetrical* stakeholder values. This means that if a program has a start, there must also be an exit. If there is a delete function, there must be also some add function. Generally speaking, the set of values must be a useful whole.

- Every subsequent delivery must show a *clear difference*. Because we want to have stakeholder feedback, the stakeholder must see a difference to feedback on. If the stakeholder feels no difference he feels that he is wasting his time and loses interest to generate feedback in the future.

- Every delivery delivers the *smallest clear increment*. If a delivery is planned, try to delete anything that is not absolutely necessary to fulfil the previous checks. If the resulting delivery takes more than two weeks, try harder.

C. Task Conclusion Criteria

If we ask different people about the contents of a defined task, all will tell a more or less different story. In order to make sure that the developer develops the right solution, we use a TaskSheet (details see [8]).

Depending on the task to be done, TaskSheets may be slightly different. First, the developer writes down on the TaskSheet:

- The requirements of the result of the task.

- Which activities must be done to complete the task.

- Design approach: how to implement it.

- Verification approach: how to make sure that it *does* what it *should do* and *does not do* what it should *not do*, based on the requirements.

- Planning (if more than one day work). If this is difficult, ask: "What am I going to do the first day."

- Anything that is not yet clear.

Then the TaskSheet is reviewed by the system architect. In this process, what the developer *thinks* has to be done is compared with what the system architect *expects*: will the result fit in the big picture? Usually there is some difference between these two views and it is better to find and resolve these differences *before* the actual execution of the task than *after*. This simply saves time.

After agreement, the developer does the work, verifies that the result produced not less, but also *not more*, than the requirements asked for. *Nice things* are not allowed: Anything not specified in the requirements is not tested. Nobody knows about it and this is an unresolvable and therefore unwanted risk.

Finally, the developer uses the task conclusion criteria on the TaskSheet to determine that the task is really done. These criteria may be adapted to certain types of tasks. In practical projects, where software code was written we used the following list:

- The code compiles and links with all files in the integration promotion level.

- The code simply does what it should do: no bugs.

- There are no memory leaks.

- Defensive programming measures have been implemented.

- All files are labelled according to the agreed rules.

- File promotion is done.

- I feel confident that the tester will find no problems.

This checklist is to make sure that the task is really done. If all checks are OK, then the work is done. If it later turns out that the work was not completely done, then the checklist is changed. Many projects where we start introducing Evo are already running. We organize an Evo-day to turn the project into an Evo project. The project has already more or less insight into what has to be done, so this can be estimated, prioritized and selected.

5. Introducing Evo in New Projects

In the case of completely new projects many team members do not yet know what has to be done, let alone how to do it. If team members have no previous Evo experience, they can hardly define tasks and thus estimation and planning of tasks is hardly possible. Still, the goal of the first Evo day is that at the end of the day the team knows roughly what to do in the coming weeks, and exactly what to do the first week. So there is a potential problem.

This problem can be solved by:

- Defining the goal of the project.

- Defining the critical success factors and the expectations of the project result from key stakeholders.

- Defining what should be done first. What do you think you should be starting on first? Like:

  - Requirements gathering.

  - Experiments.

  - Collecting information about possible tools, languages, environments.

  - Getting to know the selected tools, languages, environments, checking whether they live up to their promise.

When we ask how much time the team members are going to spend on these activities, the answer is usually "I don't know," "I don't know what I am going to search for, what I am going to find or going to decide, so I cannot estimate." This may be true, but should not be used as a licence to spend time freely. Define important things that should be done. Define time boxes, like "Use 10 hours on Requirements collection," or "Use 8 hours to draw up a tool inventory." Then put these tasks on the list of Candidate Tasks, define priorities and let everybody take 26 hours of the top of the list, get commitments and that's it. Then, the next week, based on the findings of the first week, the team is already getting a better idea of what really has to be done. The "fuzzy front end" of projects usually eats up a lot of project time, because the team lacks focus in defining what really has to be done in the project. Evo helps to keep focus and to learn quickly, by evolutionary iterations, what the project is really about.

Still, in some cases the team members cannot set their mind to commit to not-so-clear tasks within a time box. Then, but only as a last resort, team members may do whatever they want, provided that during the work they record what they are doing and how long. This is learning material for the next weeks' meeting. Note that especially if the task is not so clear, it is better first to make it clearer, before spending too much time on it.

These problems can be avoided if we start the new project with people who already have worked the Evo way. Then they know why and how to define tasks, define time boxes, set priorities and finish tasks. This enables them to start any project efficiently, without constantly asking why it has to be done this way. They *know* why and how. Starting using Evo at a completely new project adds two challenges: learning what the project is all about and learning Evo. It is easier to start learning Evo on a running project, because then the project is already known and only Evo has to be added. However, if there is no Evo experience available when starting a new project, it is still advisable to start using Evo even then, simply because it will lead to better results faster. In this case, a good coach is needed even more to make Evo succeed the first time.

6. Testing With Evo

When developing the conventional way, testing is done at the end of the development, after the Big Bang delivery. Testers then tend to find hundreds of defects, which take a long time to repair. And because there are so many defects, these tend to influence each other. Besides, repairing defects causes more defects.

Software developers are not used to using statistics. If we agree that testing never covers 100% of the software, this means that testing is *taking a sample*. At school we learnt that if we sample, we should use statistics to say something about the whole. So we should get used to statistics and not run away from it.

Statistics tell us that testing is on average 50% effective. Until you have your own (better?) figures, we have to stick to this figure. This means that the user will find the same amount of defects as found in the test. Paradoxically this means that the more defects we find in test, the more the user will find. Or, if we do not want the user to find any defects, the test should find no defects *at all*. Most developers think that defect-free software is impossible. If we extrapolate this, it means that we think it is quite normal that our car may stop after a few kilometers drive. Or that the steering wheel in some cases works just the other way: the car turns to the left when we steered to the right… Is that normal?

In Evo, we expect the developers to deliver zero-defect results for the final validation, so that the testers just have to check that everything works OK, as required. Although software developers usually start laughing by this very idea, we are very serious about this. The aim of testing earlier deliveries of Evo cycles is not just testing whether it "works". Also, testing is not to make life difficult for the developers. In Evo, the software developers ask the testers to help them to find out how far the developers are from the capability of delivering a defect free product at, or before, final validation (Figure 6-11).

Figure 6-11: Testing of early deliveries helps the developers to get ready for zero-defect final delivery.

7. Change Requests and Problem Reports

Change Requests (CR) are requested changes in the requirements. Problems Reports (PR) report things found wrong (defects), which we should have done right in the first place. Newly Defined Tasks (NT) are tasks we forgot to define. If any of these is encountered, we never start just changing, repairing, or doing the new task. We work only on defined tasks, of which the effort has been estimated and the priority defined. All tasks are listed on the list of candidate tasks in order of priority. Any CR, PR or NT is first collected in a database. This could be anything between a real database application and a notebook. Regularly, the database is analyzed by a Change Control Board (CCB). This could be anything between a very formal group of selected people, who can and must analyze the issues (CRs, PRs and NTs), and an informal group of e.g. the project manager and a team member, who check the database and decide what to do. The CCB can decide to ignore or postpone some issues, to define a new task immediately or to define an analysis task first (Figure 6-12). In an analysis task, the consequences of the issue are first analyzed and an advice is documented about what to do and what the implications are. Any task generated in this process is put on the list of candidate tasks, estimated and prioritized. And only when an existing or new task appears at the top of the candidate list will it be worked on.

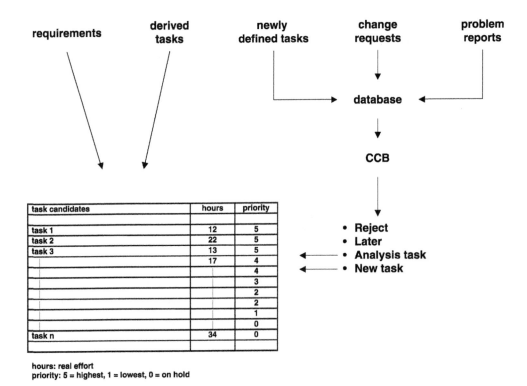

hours: real effort
priority: 5 = highest, 1 = lowest, 0 = on hold

Figure 6-12: All activities, including Change Requests, Problem Reports and Newly Defined Tasks use the same mechanism for estimation and prioritizing: the list of candidate tasks.

8. Tools

Special tools may only be used when we know and understand the right methods. In actual projects, we have used MS Excel as an easy notepad during interactive sessions with a LCD projector showing what happens on this notepad in real time. When tasks have been defined, MS Project can be used as a spreadsheet to keep track of the tasks per person, while automatically generating a time-line in the Gantt-chart view (Figure 6-13, top left). This time-line tells people, including management, more than textual planning. It proved possible to let MS Project use weeks of 26 hours and days of 5.2 hours, so that durations could be entered in real effort while the time-line shows correct lead time-days.

There is a relationship between requirements, stakeholder values, deliveries and tasks (Figure 6-13). We even want to have different views on the list of tasks, like a list of prioritized candidate tasks of the whole project and lists of prioritized tasks per developer. This calls for the use of a relational database, to organize the relations between requirements, values, deliveries and tasks and the different views. Currently, such a database has not been made and the project manager has to keep the consistency of the relations manually. This is some extra work. However, in the beginning it helps the project manager knowing what he is doing. And when we will have found the best way to do it and found the required relationships and views we really need, then we could specify the requirements of a database. If we would have

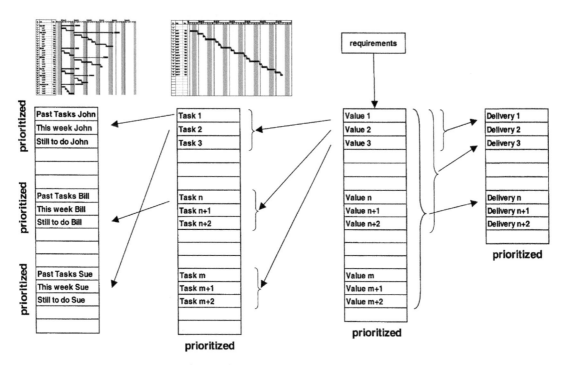

Figure 6-13: Relations between requirements, stakeholder values, deliveries, and different views on tasks.

waited till we had a database to keep track of all things, we probably would not have started gaining Evo experience yet. Whether existing tools, such as from Rational, can solve this database problem sufficiently, is interesting to investigate.

Important before selecting any tool is, however, to know what we want to accomplish and why and how. Only then can we check whether the tool could save time and bureaucracy rather than costing time and bureaucracy.

9. Conclusion

We described issues that are addressed by the Evo methods and the way we organize Evo projects. By using these methods in actual projects we find:

- *Faster results*
  Evo projects deliver better results in 30% shorter time than otherwise. Note: 30% shorter than what by conventional methods would have been achieved. This may be longer than initially hoped for. Although this 30% is not scientifically proven, it is rather plausible by considering that we constantly check whether we are doing the *right things in the right order to the right level of detail for that moment.* This means that any other process is always less efficient. Most processes (even if you don't know which process you follow, you are following an intuitive *ad hoc* process) cause much work to be done incorrectly and then repaired, as well as unnecessary work. Most developers admit that they use more than half of the total project time on debugging. That is repairing things they did wrong the first time. In Evo, most "bugs" are prevented.

- *Better quality*
  We define quality as (Crosby [4]) "Conformance to Requirements" (How else can we design for quality and measure quality?). In Evo we constantly reconsider the validity of the requirements and our assumptions and make sure that we deliver the most important requirements first. Thus the result will be at least as good as what is delivered with the less rigorous approach we encounter in other approaches.

- *Less stressed developers*
  In conventional projects, where it is normal that tasks are not completed in time, developers constantly feel that they fail. This is very demotivating. In Evo projects, developers succeed regularly and see regularly real results of their work. People enjoy success. It motivates greatly. And because motivation is the motor of productivity, the productivity soars. This is what we see happening *within two weeks* in Evo projects: People get relaxed, happy, smiling again, while producing more.

- *Happy customers*
  Customers enjoy getting early deliveries and producing regular feedback. They know that they have difficulty in specifying what they really need. By showing them early deliveries and being responsive to their requirements changes, they feel that we know what we are doing. In other developments, they are constantly anxious about the

result, which they get only at the end, while experience tells them that the first results are usually not OK and too late. Now they get actual results even much earlier. They start trusting our predictions. And they get a choice of time to market because we deliver complete, functioning results, with growing completeness of functions and qualities, well before the deadline. This has never happened before.

■ *More profits*
If we use less time to deliver better quality in a predictable way, we save a lot of money, while we can earn more money with the result. Combined, we make a lot more profit.

In short, although Brooks predicted a long time ago that "There is no silver bullet" [2], we found that the methods presented, which are based on ideas practiced even before the "silver bullet" article, do seem to be a "magic bullet" because of the remarkable results obtained.

Acknowledgment

A lot (but not all) of the experience with the approach described in this booklet has originally been gained at Philips Remote Control Systems, Leuven, Belgium.

In a symbiotic cooperation with the group leader, Bart Vanderbeke, the approach has been introduced in all software projects of his team. Using short discuss-implement-check-act improvement cycles during a period of 8 months, the approach led to a visibly better manageability and an increased comfort-level for the team members, as well as for the product managers.

We would like to thank the team members and product managers for their contribution to the results.

References

1. H.D. Mills: *Top-Down Programming in Large Systems. In Debugging Techniques in Large Systems.* Ed. R. Ruskin, Englewood Cliffs, NJ: Prentice Hall, 1971.

2. F.P. Brooks, Jr.: *No Silver Bullet: Essence and Accidents of Software Engineering.* In Computer vol 20, no.4 (April 1987): 10–19.

3. T. Gilb: *Principles of Software Engineering Management.* Addison-Wesley Pub Co, 1988, ISBN: 0201192462.

4. P.B. Crosby: *Quality Is Still Free.* McGraw-Hill, 1996. 4th edition ISBN 0070145326

5. T. Gilb: manuscript: *Evo: The Evolutionary Project Managers Handbook.* http://www. gilb.com/pages/2ndLevel/gilbdownload.html, 1997.

6. S.J. Prowell, C.J. Trammell, R.C. Linger, J.H.Poore: *Cleanroom Software Engineering, Technology and Process.* Addison-Wesley, 1999, ISBN 0201854805.

7. T. Gilb: manuscript: *Impact Estimation Tables: Understanding Complex Technology Quantatively*. http://www.gilb.com/pages/2ndLevel/gilbdownload.html 1997.

8. N.R. Malotaux: TaskSheet. http://www.malotaux.nl/nrm/English/Forms.htm, 2000.

9. F.P. Brooks, Jr.: *The Mythical Man-month*. Addison-Wesley, 1975, ISBN 0201006502. Reprint 1995, ISBN 0201835959.

10. C. Northcote Parkinson: *Parkinsons Law*. Buccaneer Books, 1996, ISBN 1568490151.

11. N.R. Malotaux: Powerpoint slides: *Evolutionary Delivery*. 2001. http://www.malotaux.nl/nrm/pdf/EvoIntro.pdf.

12. E. Dijkstra: Paper: *Programming Considered as a Human Activity*, 1965. Reprint in *Classics in Software Engineering*. Yourdon Press, 1979, ISBN 0917072146.

13. W. A. Shewhart: *Statistical Method from the Viewpoint of Quality Control*. Dover Publications, 1986. ISBN 0486652327.

14. W.E. Deming: *Out of the Crisis*. MIT, 1986, ISBN 0911379010.

15. Kent Beck: *Extreme Programming Explained*, Addison Wesley, 1999, ISBN 0201616416.

16. http://www.gilb.com.

17. http://www.extremeprogramming.org.

Embedded State Machine Implementation

Turning a state machine into a program
can be straightforward if you follow
the advice of a skilled practitioner.

Martin Gomez

Martin Gomez is a software engineer at the Johns Hopkins University Applied Physics Lab, where he is presently developing flight software for a solar research spacecraft. He has been working in the field of embedded software development for 17 years. Martin has a BS in aerospace engineering and an MS in electrical engineering, both from Cornell University. He may be reached at martin.gomez@jhuapl.edu.

Many embedded software applications are natural candidates for mechanization as a state machine. A program that must sequence a series of actions, or handle inputs differently depending on what mode it's in, is often best implemented as a state machine.

This article describes a simple approach to implementing a state machine for an embedded system. Over the last 15 years, I have used this approach to design dozens of systems, including a softkey-based user interface, several communications protocols, a silicon-wafer transport mechanism, an unmanned air vehicle's lost-uplink handler, and an orbital mechanics simulator.

State Machines

For purposes of this article, a state machine is defined as an algorithm that can be in one of a small number of states. A state is a condition that causes a prescribed relationship of inputs to outputs, and of inputs to next states. A savvy reader will quickly note that the state machines described in this article are Mealy machines. A Mealy machine is a state machine where the outputs are a function of both present state and input, as opposed to a Moore machine, in which the outputs are a function only of state. [1] In both cases, the next state is a function of both present state and input. Pressman has several examples of state transition diagrams used to document the design of a software product.

Figure 7-1 shows a state machine. In this example, the first occurrence of a slash produces no output, but causes the machine to advance to the second state. If it encounters a non-slash

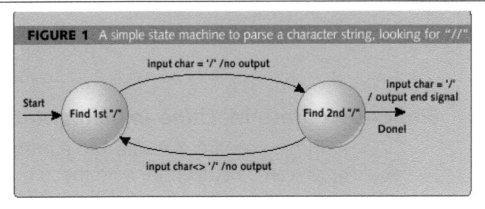

FIGURE 1 A simple state machine to parse a character string, looking for "//"

input char = '/' /no output

Start

Find 1st "/"

input char = '/'
/ output end signal

Find 2nd "/"

Done!

input char<> '/' /no output

Figure 7-1: A simple state machine to parse a character string looking for "//."

while in the second state, then it will go back to the first state, because the two slashes must be adjacent. If it finds a second slash, however, then it produces the "we're done" output.

The state machine approach I recommend proceeds as follows:

- Learn what the user wants
- Sketch the state transition diagram
- Code the skeleton of the state machine, without filling in the details of the transition actions
- Make sure the transitions work properly
- Flesh out the transition details
- Test

An Example

A more illustrative example is a program that controls the retraction and extension of an airplane's landing gear. While in most airplanes this is done with an electrohydraulic control mechanism (simply because they don't have a computer on board), cases exist—such as unmanned air vehicles—where one would implement the control mechanism in software.

Let's describe the hardware in our example so that we can later define the software that controls it. The landing gear on this airplane consists of a nose gear, a left main gear, and a right main gear. These are hydraulically actuated. An electrically driven hydraulic pump supplies pressure to the hydraulic actuators. Our software can turn the pump on and off. A direction valve is set by the computer to either "up" or "down," to allow the hydraulic pressure to either raise or lower the landing gear. Each leg of the gear has two limit switches: one that closes if the gear is up, and another that closes when it's locked in the down position. To determine if the airplane is on the ground, a limit switch on the nose gear strut will close if the weight of the airplane is on the nose gear (commonly referred to as a "squat switch"). The pilot's controls consist of a landing gear up/down lever and three lights (one per leg) that can either be off, glow green (for down), or glow red (for in transit).

Let us now design the state machine. The first step, and the hardest, is to figure out what the user really wants the software to do. One of the advantages of a state machine is that it forces the programmer to think of all the cases and, therefore, to extract all the required information from the user. Why do I describe this as the hardest step? How many times have you been given a one-line problem description similar to this one: don't retract the gear if the airplane is on the ground.

Clearly, that's important, but the user thinks he's done. What about all the other cases? Is it okay to retract the gear the instant the airplane leaves the ground? What if it simply bounced a bit due to a bump in the runway? What if the pilot moved the gear lever into the "up" position while he was parked, and subsequently takes off? Should the landing gear then come up?

One of the advantages of thinking in state machine terms is that you can quickly draw a state transition diagram on a whiteboard, in front of the user, and walk him through it. A common notation designates state transitions as follows: < event that caused the transition >/< output as a result of the transition >. [2] If we simply designed what the user initially asked us for ("don't retract the gear if the airplane is on the ground"), what he'd get would look a bit like Figure 7-2. It would exhibit the "bad" behavior mentioned previously.

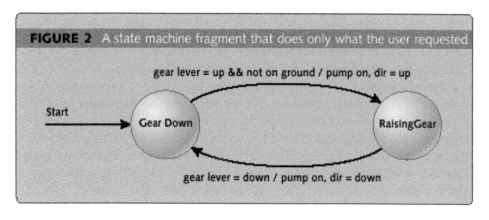

Figure 7-2: A state machine fragment that does only what the user requested.

Keep the following in mind when designing the state transition diagram (or indeed any embedded algorithm):

- Computers are very fast compared to mechanical hardware—you may have to wait.

- The mechanical engineer who's describing what he wants probably doesn't know as much about computers or algorithms as you do. Good thing, too—otherwise you would be unnecessary!

- How will your program behave if a mechanical or electrical part breaks? Provide for timeouts, sanity checks, and so on.

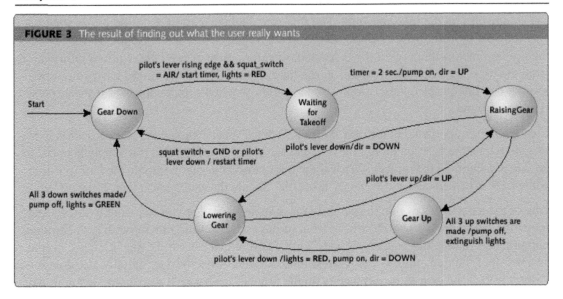

FIGURE 3 The result of finding out what the user really wants

Figure 7-3: The result of finding out what the user really wants.

We can now suggest the following state machine to the user, building upon his requirements by adding a few states and transitions at a time. The result is shown in Figure 7-3. Here, we want to preclude gear retraction until the airplane is definitely airborne, by waiting a couple of seconds after the squat switch opens. We also want to respond to a rising edge of the pilot's lever, rather than a level, so that we rule out the "someone moved the lever while the airplane was parked" problem. Also, we take into account that the pilot might change his mind. Remember, the landing gear takes a few seconds to retract or extend, and we have to handle the case where the pilot reversed the lever during the process. Note, too, that if the airplane touches down again while we're in the "Waiting for takeoff" state, the timer restarts—the airplane has to be airborne for two seconds before we'll retract the gear.

Implementation

This is a good point to introduce a clean way to code a finite state machine. Listing 7-1 is my implementation of the state machine in Figure 7-3.

Listing 7-1: Landing gear implementation.

```c
typedef enum {GEAR_DOWN = 0, WTG_FOR_TKOFF, RAISING_GEAR, GEAR_UP,
LOWERING_GEAR} State_Type;

/* This table contains a pointer to the function to call in each state.*/
void (*state_table[]090 = {GearDown, WtgForTakeoff, RaisingGear, GearUp,
LoweringGear};

State_Type curr_state;

Main()
{
    InitializeLdgGearSM();

    /*  The heart of the state machine is this one loop. The function
        corresponding to the current state is called once per iteration. */
    while (1)
    {
        state_table[curr_state]();
        DecrementTimer();

        /* Do other functions, not related to this state machine.*/
    }
};

void InitializeLdgGearSM()
{
    curr_state = GEAR_DOWN;
    timer = 0.0;

    /* Stop all the hardware, turn off the lights, etc.*/
}

void GearDown()
{
    /* Raise the gear upon command, but not if the airplane is on the
    ground.*/
    if ((gear_lever == UP) && (prev_gear_lever == DOWN) && (squat_switch ==
    UP))
    {
        timer = 2.0;
        curr_state = WTG_FOR_TKOFF;
    };

    prev_gear_lever = gear_lever; /* Store for edge detection.*/
}

void RaisingGear()

{
    /* Once all 3 legs are up, go to the GEAR_UP state.*/
    if ((nosegear_is_up == MADE) && (leftgear_is_up == MADE) &&
    (rtgear_is_up    == MADE))
    {
```

Listing 7-1 continued on next page

Listing 7-1: Landing gear implementation (continued).

```
        Curr_state = GEAR_UP;
    };

    /* If the pilot changes his mind, start lowering the gear.*/
    if (gear_lever == DOWN)
    {
        curr_state = LOWERING_GEAR;
    };
}

void GearUp()
{
    /* If the pilot moves the lever to DOWN, lower the gear.*/
    if (gear_lever == DOWN)
    {
        curr_state = LOWERING_GEAR;
    };
}

void WtgForTakeoff()
{
    /* Once we've been airborne for 2 sec., start raising the gear.*/
    if (timer <=0.0)
    {
        curr_state = RAISING_GEAR;
    };

    /*If we touch down again, or if the pilot changes his mind, start
over.*/
    if ((squat_switch ==DOWN) || (gear_lever == DOWN))
    {
        timer = 2.0;
        curr_state = GEAR_DOWN;

        /* Don't want to require that he toggle the lever again
           this was just a bounce.*/
        prev_gear_lever = DOWN;
    };
}

void LoweringGear()
{
    if (gear_lever == UP)
    {
        curr_state = RAISING_GEAR;
    };

    if ((nosegear_is_down == MADE) && (Leftgear_is_down == MADE) &&
        (rtgear_is_down == MADE))
    {
        curr_state = GEAR_DOWN;
    };
}
```

Let's discuss a few features of the example code. First, you'll notice that the functionality of each individual state is implemented by its own C function. You could just as easily implement it as a switch statement, with a separate case for each state. However, this can lead to a very long function (imagine 10 or 20 lines of code per state for each of 20 or 30 states.) It can also lead you astray when you change the code late in the testing phase—perhaps you've never forgotten a break statement at the end of a case, but I sure have. Having one state's code "fall into" the next state's code is usually a no-no.

To avoid the switch statement, you can use an array of pointers to the individual state functions. The index into the array is curr_state, which is declared as an enumerated type to help our tools enforce correctness.

In coding a state machine, try to preserve its greatest strength, namely, the eloquently visible match between the user's requirements and the code. It may be necessary to hide hardware details in another layer of functions, for instance, to keep the state machine's code looking as much as possible like the state transition table and the state transition diagram. That symmetry helps prevent mistakes, and is the reason why state machines are such an important part of the embedded software engineer's arsenal. Sure, you could do the same thing by setting flags and having countless nested if statements, but it would be much harder to look at the code and compare it to what the user wants. The code fragment in Listing 7-2 fleshes out the RaisingGear() function.

Listing 7-2: The RaisingGear() function.

```
Void RaisingGear()
{
    /* Once all 3 legs are up, go to the GEAR_UP state.*/
    if (nosegear_is up == MADE) && (leftgear_is_up == MADE) && (rtgear_is_up
       == MADE))
    {
        pump_motor = OFF;
        gear_lights = EXTINGUISH;
        curr_state = GEAR_UP;
    };

    /* If the pilot changes his mind, start lowering the gear.*/
    if (gear_lever == DOWN)
    {
        pump_direction = DOWN;
        curr_state = GEAR_LOWERING;
    };
}
```

Notice that the code for RaisingGear() attempts to mirror the two rows in the state transition table for the Raising Gear state.

As an exercise, you may want to expand the state machine we've described to add a timeout to the extension or retraction cycle, because our mechanical engineer doesn't want the hydraulic pump to run for more than 60 seconds. If the cycle times out, the pilot should be alerted by alternating green and red lights, and he should be able to cycle the lever to try again. Another feature to exercise your skills would be to ask our hypothetical mechanical engineer "Does the pump suffer from having the direction reversed while it's running? We do it in the two cases where the pilot changes his mind." He'll say "yes," of course. How would you modify the state machine to stop the pump briefly when the direction is forced to reverse?

Testing

The beauty of coding even simple algorithms as state machines is that the test plan almost writes itself. All you have to do is to go through every state transition. I usually do it with a highlighter in hand, crossing off the arrows on the state transition diagram as they successfully pass their tests. This is a good reason to avoid "hidden states"—they're more likely to escape testing than explicit states. Until you can use the "real" hardware to induce state changes, either do it with a source-level debugger, or build an "input poker" utility that lets you write the values of the inputs into your application.

This requires a fair amount of patience and coffee, because even a mid-size state machine can have 100 different transitions. However, the number of transitions is an excellent measure of the system's complexity. The complexity is driven by the user's requirements: the state machine makes it blindingly obvious how much you have to test. With a less-organized approach, the amount of testing required might be equally large—you just won't know it.

It is very handy to include print statements that output the current state, the value of the inputs, and the value of the outputs each time through the loop. This lets you easily observe what ought to be the Golden Rule of Software Testing: don't just check that it does what you want—also check that it doesn't do what you don't want. In other words, are you getting only the outputs that you expect? It's easy to verify that you get the outputs that you expected, but what else is happening? Are there "glitch" state transitions, that is, states that are passed through inadvertently, for only one cycle of the loop? Are any outputs changing when you didn't expect them to? Ideally, the output of your prints would look a lot like the state transition table.

Finally, and this applies to all embedded software and not just to that based on state machines, be suspicious when you connect your software to the actual hardware for the first time. It's very easy to get the polarity wrong—"Oh, I thought a '1' meant raise the gear and a '0' meant lower the gear." On many occasions, my hardware counterpart inserted a temporary "chicken switch" to protect his precious components until he was sure my software wasn't going to move things the wrong way.

Crank It

Once the user's requirements are fleshed out, I can crank out a state machine of this complexity in a couple of days. They almost always do what I want them to do. The hard part, of course, is making sure that I understand what the user wants, and ensuring that the user knows what he wants—that takes considerably longer!

References

1. Hurst, S.L. *The Logical Processing of Digital Signals*. New York: Crane, Russak, 1978.

2. Pressman, Roger A. *Software Engineering: A Practitioner's Approach*, 3rd Edition. New York: McGraw-Hill, 1992.

3. Shumate, Kenneth C. and Marilyn M. Keller. *Software Specification and Design: A Disciplined Approach for Real-Time Systems*. New York: John Wiley & Sons, 1992.

Hierarchical State Machines

Sandeep Ahluwalia

Sandeep Ahluwalia is a writer and developer for EventHelix.com Inc., a company that is dedicated to developing tools and techniques for real-time and embedded systems. They are developers of the EventStudio, a CASE tool for distributed system design in object-oriented as well as structured development environments. They are based in Gaithersburg, Maryland. Their website is www.eventhelix.com.

In this chapter, we will be highlighting the advantages of hierarchical state machine design over conventional state machine design.

In conventional state machine design, all states are considered at the same level. The design does not capture the commonality that exists among states. In real life, many states handle most messages in a similar fashion and differ only in handling of a few key messages. Even when the actual handling differs, there is still some commonality. Hierarchical state machine design captures the commonality by organizing the states as a hierarchy. The states at the higher level in hierarchy perform the common message handling, while the lower level states inherit the commonality from higher level ones and perform the state specific functions. The table given below shows the mapping between conventional states and their hierarchical counterparts for a typical call state machine.

Conventional States	Hierarchical States
Awaiting First Digit	Setup.CollectingDigits.AwaitingFirstDigit
Collecting Digits	Setup.CollectingDigits.AwaitingSubsequent Digits
Routing Call	Setup.RoutingCall
Switching Path	Setup.SwitchingPath
Conversation	Conversation
Awaiting Onhook	Releasing.AwaitingOnhook
Releasing Path	Releasing.ReleasingPath

A conventional state machine is designed as a two dimensional array with one dimension as the state and the other dimension specifying the message to be handled. The state machine determines the message handler to be called by indexing with the current state and the received message. In real life scenario, a task usually has a number of states along with many different types of input messages. This leads to a message handler code explosion. Also, a huge two dimensional array needs to be maintained. Hierarchical state machine design avoids this problem by recognizing that most states differ in the handling of only a few messages. When a new hierarchical state is defined, only the state specific handlers need to be specified.

Conventional State Machine Example

Figure 8-1 describes the state transition diagram for an active standby pair. The design here assumes that the active and standby are being managed by an external entity.

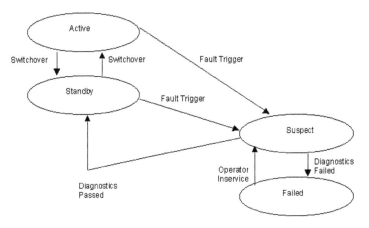

Figure 8-1.

The different states for the state machine are Active, Standby, Suspect and Failed. The input messages to be handled are Switchover, Fault Trigger, Diagnostics Passed, Diagnostics Failed and Operator Inservice. Thus the handler two dimensional array is 4 × 5, i.e., 20 handlers need to be managed.

Listing 8-1 shows the handlers that need to be defined. A dummy "do nothing" handler should be specified for all other entries of the two dimensional state table. This simple example clearly illustrates the problem with conventional state design. There is a lot of code repetition between handlers. This creates a maintenance headache for state machine designers. We will see in the following section that hierarchical state machine design exploits these very similarities to implement a more elegant state structure.

Listing 8-1: Conventional state machine implementation.

```
/* == Active State Handlers == */
void ActiveStateFaultTriggerHandler(Msg *pMsg)
  {
      PerformSwitchover();              // Perform Switchover, as active failed
      NextState = SUSPECT;             // Run diagnostics to confirm fault
      SendDiagnosticsRequest();
      RaiseAlarm(LOSS_OF_REDUNDANCY); // Report loss of redundancy to
                                              operator
  }

void ActiveStateSwitchoverHandler(Msg *pMsg)
{
  PerformSwitchover();                // Perform Switchover on operator command
  CheckMateStatus();                  // Check if switchover completed
  SendSwitchoverResponse();           // Inform operator about switchover
  NextState = STANDBY;                // Transition to standby
}

/* == Standby State Handlers == */
void StandbyStateFaultTriggerHandler(Msg *pMsg)
{
  NextState = SUSPECT;                // Run diagnostics to confirm fault
  SendDiagnosticsRequest();
  RaiseAlarm(LOSS_OF_REDUNDANCY);    // Report loss of redundancy to
                                              operator
}

void StandbyStateSwitchoverHandler(Msg *pMsg)
{
  PerformSwitchover();                // Perform switchover on operator command
  CheckMateStatus();                  // Check if switchover completed
  SendSwitchoverResponse();           // Inform operator about switchover
  NextState = ACTIVE;                 // Transition to active
}

/* == Suspect State Handlers == */
void SuspectStateDiagnosticsFailedHandler(Msg *pMsg)
{
  SendDiagnosticsFailureReport();    // Inform operator about diagnostics
  NextState = FAILED;                // Move to the failed state }

void SuspectStateDiagnosticsPassedHandler(Msg *pMsg)
{
  SendDiagnosticsPassReport();       // Inform operator about diagnostics
  ClearAlarm(LOSS_OF_REDUNDANCY);    // Clear loss of redundancy alarm
  NextState = STANDBY;               // Move to standby state
}


(Listing 8-1 continued on next page)
```

Listing 8-1: Conventional state machine implementation (continued).

```
void SuspectStateOperatorInservice(Msg *pMsg)
{
  // Operator has replaced the card, so abort the current diagnostics
  // and restart new diagnostics on the replaced card.
  AbortDiagostics();
  SendDiagnosticsRequest();          // Run diagnostics on replaced card
  SendOperatorInserviceResponse();   // Inform operator about diagnostics
                                     start

}
/* == Failed State Handlers == */
void FailedStateOperatorInservice(Msg *pMsg)
{
  SendDiagnosticsRequest();          // Run diagnostics on replaced card
  SendOperatorInserviceResponse();   // Inform operator about diagnostics
                                     start
  NextState = SUSPECT;               // Move to suspect state for diagnostics
}
```

Hierarchical State Machine Example

The following state transition diagram recasts the state machine by introducing two levels in the hierarchy. Inservice and OutOfService are the high level states that capture the common message handling. Active and Standby states are low level states inheriting from Inservice state. Suspect and Failed are low level states inheriting from the OutOfService state.

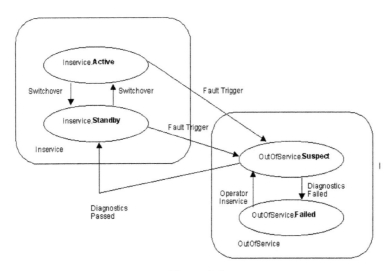

Figure 8-2.

The following diagram clearly illustrates the state hierarchy. Even the Inservice and OutOfService high level states inherit from the UnitState that is at the highest level.

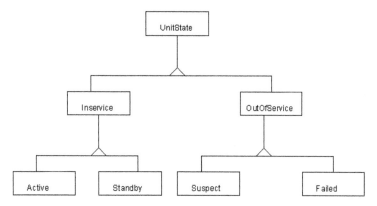

Figure 8-3.

The following code declares all the hierarchical states. The highest level state defines all the dummy "do nothing" handlers, thus the inheriting lower level states need not worry about messages that they do not handle. All the handlers have been declared virtual so that they can be overridden by the deriving states.

The code also covers the implementation of the Unit class state machine. The states are defined as static objects in the Unit class. Thus, they are shared amongst all the Unit state machine instances. The state machine also maintains the current state pointer which is used to invoke the appropriate handler in the OnMessage method. The state transitions are carried out by calling the NextState method. NextState marks the specified state as the current state.

Listing 8-2: Hierarchical state machine groundwork.

```
// Base class for all states of the Unit state machine. Default handlers
// are defined for all the events
class UnitState
{
public:
    virtual void OnSwitchover(Unit &u, Msg *pMsg)  {}
    virtual void OnFaultTrigger(Unit &u, Msg *pMsg) {}
    virtual void OnDiagnosticsFailed(Unit &u, Msg *pMsg) {}
    virtual void OnDiagnosticsPassed(Unit &u, Msg *pMsg) {}
    virtual void OnOperatorInservice(Unit &u, Msg *pMsg) {}
};

// Base class for all Inservice states. Common handling for switchover

// and Fault Trigger messages is defined.

(Listing 8-2 continued on next page)
```

Listing 8-2: Hierarchical state machine groundwork (continued).

```cpp
class Inservice : public UnitState
{
public:
    void OnSwitchover(Unit &u, Msg *pMsg);
    void OnFaultTrigger(Unit &u, Msg *pMsg);
};

// The Active state inherits from Inservice state and further
// refines the Switchover and Fault Trigger handling

class Active : public Inservice
{
public:
    void OnSwitchover(Unit &u, Msg *pMsg);
    void OnFaultTrigger(Unit &u, Msg *pMsg);
};

// The Standby state inherits from the Inservice state and further
// refines the Switchover handler

class Standby : public Inservice
{ public:
    void OnSwitchover(Unit &u, Msg *pMsg);
};

// Base class for all out of service classes. This class acts as
// the base class for all out of service states.

class OutOfService : public UnitState
{
public:
    void OnOperatorInservice(Unit &u, Msg *pMsg);
};

class Suspect : public OutOfService
{
public:
    void OnDiagnosticsFailed(Unit &u, Msg *pMsg);
    void OnDiagnosticsPassed(Unit &u, Msg *pMsg);
};

class Failed : public OutOfService
{
public:
    // No Need to Override any other method
};
class Unit
{
public:
    static Active ActiveState;
    static Standby StandbyState;
    static Suspect SuspectState;
    static Failed FailedState;
```

(**Listing 8-2** continued on next page)

Listing 8-2: Hierarchical state machine groundwork (continued).

```cpp
    void OnMessage(Msg *pMsg);
    void NextState(UnitState &rState);

    // Common Methods invoked from several states
    // (See article on FSM Inheritance for details)
    virtual void SendDiagnosticsRequest();
    virtual void RaiseAlarm(int reason);
    virtual void ClearAlarm(int reason);
    virtual void PerformSwitchover();
    . . .
    virtual void SendSwitchoverResponse();
    virtual void AbortDiagnostics();
private:
    UnitState *pCurrentState;
};

void Unit::NextState(UnitState &rState)
{
    pCurrentState = &rState;
}

void Unit::OnMessage(Msg *pMsg)
{
    switch (pMsg->getType())
    {
    case FAULT_TRIGGER:
        pCurrentState->OnFaultTrigger(*this, pMsg);
        break;

    case SWITCHOVER:
        pCurrentState->OnSwitchover(*this, pMsg);
        break;

    case DIAGNOSTICS_PASSED:
        pCurrentState->OnDiagnosticsPassed(*this, pMsg);
        break;

    case DIAGNOSTICS_FAILED:
        pCurrentState->OnDiagnosticsFailed(*this, pMsg);
        break;

    case OPERATOR_INSERVICE:
        pCurrentState->OnOperatorInservice(*this, pMsg);
        break;

    default:
        assert(FALSE);
        break;
    }
}
```

The implementation details of the various state handler methods are given in Listing 8-3. It is apparent that all the commonality has moved to the high level states, viz. Inservice and OutOfService. Also, contrast this with the conventional state machine implementation.

Listing 8-3: Hierarchical state machine implementation.

```
/* == Inservice State Handlers == */

void Inservice::OnFaultTrigger(Unit &u, Msg *pMsg)
{
    u.NextState(&u.SuspectState);
    u.SendDiagnosticsRequest();
    u.RaiseAlarm(LOSS_OF_REDUNDANCY);
}

void Inservice::OnSwitchover(Unit &u, Msg *pMsg)
{
  u.PerformSwitchover();
  u.CheckMateStatus();
  u.SendSwitchoverResponse();
}

/* == Active State Handlers == */

void Active::OnFaultTrigger(Unit &u, Msg *pMsg)
  {
    u.PerformSwitchover();
    Inservice::OnFaultTrigger(u, pMsg);
}

void Active::OnSwitchover(Unit &u, Msg *pMsg)
{
  Inservice::OnSwitchover(u, pMsg);
  u.NextState(u.StandbyState);
}

/* == Standby State Handlers == */

void Standby::OnSwitchover(Unit &u, Msg *pMsg)
{
  Inservice::OnSwitchover(u, pMsg);
  u.NextState(u.ActiveState);
}

/* == OutOfService State Handlers == */

void OutOfService::OperatorInservice(Unit &u, Msg *pMsg)
{
    // Operator has replaced the card, so abort the current diagnostics
    // and restart new diagnostics on the replaced card.
    u.SendDiagnosticsRequest();
    u.SendOperatorInserviceResponse();
    u.NextState(u.SuspectState);
}

(Listing 8-3 continued on next page)
```

Listing 8-3: Hierarchical state machine implementation (continued).

```
/* == Suspect State Handlers == */
void Suspect::OnDiagnosticsFailed(Unit &u, Msg *pMsg)
{
    u.SendDiagnosticsFailureReport();
    u.NextState(u.FailedState);
}

void Suspect::OnDiagnosticsPassed(Unit &u, Msg *pMsg)
{
    u.SendDiagnosticsPassReport();
    u.ClearAlarm(LOSS_OF_REDUNDANCY);
    u.NextState(u.StandbyState);
}

void Suspect::OperatorInservice(Unit &u, Msg *pMsg)
{
    u.AbortDiagostics();
    OutOfService::OperatorInservice(u, pMsg);
}
```

Developing Safety Critical Applications

Salah Obeid

Salah Obeid is VP of Engineering at Embedded*Plus* Engineering, a system and software engineering development firm in Tempe, Arizona. He is recognized by the software and technical industry as an expert for providing training on object oriented techniques and application of C++/UML (including applicable validation/verification techniques) as they apply to safety-critical systems and software. At Embedded*Plus* Engineering, Salah provides leadership to teams working with companies in the aerospace, military, tele-communication and medical device industries. He provides support in defining system requirements, system design, simulation/modeling strategies, test definition, and the generation of software development standards. This includes the implementation and interpretation of DO-178B. He was requested by the FAA to co-chair an industry sub-team on object-oriented validation/verification tools. He is currently a member of the RTCA SC-190 committee, the FAA/NASA Object Oriented Technology in Aviation (OOTiA) committee and a member of the Object Management Group (OMG) working on the definition of UML and CORBA. Besides Embedded*Plus* Engineering, Salah has held leadership positions at IBM, Allied Signal and Honeywell. Salah can be reached at salah@embeddedplus.com or www.embeddedplus.com.

Introduction

Software plays a major role in safety critical applications from fly-by-wire airplanes, automobile computers, and cruise ship controls to controlling power plants and medical devices. Software can be at various levels of criticality. The FAA uses the DO-178B Guidelines for Safety Critical Applications to determine the criticality level and provide guidance to the software developers through out the software development process. DO-178B is currently being adapted by other industries for use in the development of safety critical software applications. This class introduces development of safety critical applications including an overview of the DO-178B process. The class also covers the use of DO-178B standard in an object-oriented iterative software development process. The advantages of using an iterative process in the development of safety-critical software will also be highlighted.

Reliability and safety

Someone may think that reliability and safety are the same, but they are not. A product may be reliable but unsafe or it may be safe but unreliable. Reliability is quite different from safety, and it's not always required in safety critical systems. For example, an MRI scanner has to be safe but not necessary reliable. Obviously if it was broken it will not be used, or will be considered to be unsafe. Most of the safety critical systems have a fail safe mode; this means that it can fail but its failure has to be safe. In our example it might be acceptable for the MRI machine to fail as long as it does not cause a safety hazard. For example, in the case where loss of function is catastrophic, reliability is necessary but not sufficient. In the case where loss of function has no effect, reliability of function availability is not necessary, and in the case where misleading data is hazardous, reliability can be an indicator.

The following diagram shows the reliability growth model, where as the fault frequency gets lower (near zero) the reliability of the system gets higher.

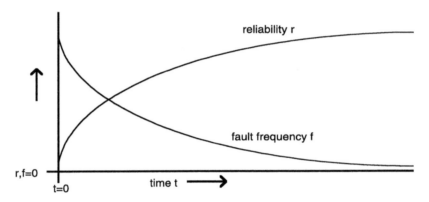

Figure 9-1: Reliability growth model.

Along came DO-178B

Because of the rapid increase in the use of software in airborne systems and equipment used on aircraft, in the early 1980s there was a need to create an industry-accepted guidance for satisfying airworthiness requirements as established by the FAA. DO-178A was created to address the need for determining with an acceptable level of confidence that the software development and certification comply with airworthiness requirements. A few years later along came DO-178B which was created to address deficiencies noted in DO-178A. DO-178B is process oriented and places additional emphasis on verification rather than just testing. It also includes focus on requirements based testing and structural coverage.

The DO-178B standard is one of the most stringent standards imposed on the software development industry. This aerospace guideline has been used to help in the development of similar standards in other fields of application such as the nuclear power, military, rail/ automotive transportation, and medical industries where reliability, dependability and safety are required in the software development process. The FAA requires Designated Engineering Representatives (DER) to oversee software development and liaise with the FAA.

Overview of DO-178B

"DO-178B—Software Considerations in Airborne Systems and Equipment Certification" is an industry standard that provides guidelines for the production of airborne systems software including tool qualification and is published by the aerospace association RTCA. The standard is used internationally to specify process requirements that ensure the safety and airworthiness of software for avionics systems. The standard also contains requirements for performing tool qualifications when a tool is used to automate software development processes that are typically performed by humans. The DO-178B guidelines describe techniques and methods appropriate to ensure that software's integrity and reliability can be ultimately certified and approved by the FAA.

The guidelines for software development and verification are described in DO-178B for each of six software criticality levels. A System Safety Assessment process is required to associate criticality levels with the various system elements. Safety is the main objective for all the processes described in the DO-178B guidelines, but the guidelines could be applied to any process utilized for other safety systems. Level A is the highest criticality and is applied when anomalous behavior causes catastrophic failure condition. Level E is the lowest level and is applied when anomalous behavior has no effect on operational capacity. For each level A through E, the guidelines provide requirements in the areas of software planning, development, verification, configuration management, quality assurance, tool qualification, certification, and maintenance.

The DO-178B guidelines categorize software levels based on their failure condition severity. Process requirements and objectives are expanded for the various failure condition severities as outlined in the guidelines. For example, structural coverage testing may or may not be required for a specific software level. A software error may cause a problem that contributes to a failure condition; therefore the level of software needed for safe operation is related to the system failure conditions.

Failure Condition Categorization

The following is a list of failure condition categories as described in DO-178B:

1. Catastrophic: Failure condition that causes loss of safe operation of the system. For example, preventing continued safe flight and landing.

2. Hazardous/Severe-Major: Failure conditions which significantly reduce the safety margins or functional capabilities of the system that could case fatal injuries.

3. Major: Failure conditions which reduce the safety margins of the system.

4. Minor: Failure conditions which would significantly reduce system (or aircraft) safety.

5. No Effect: Failure conditions which do not affect the operation capability of the system or aircraft.

The software level is determined by the system safety assessment process and is based on the contribution of software to potential system failures. The following table lists the software level and their categorization and the probability of failure:

Table 9-1: DO-178B software levels and their categorization.

Categorization	Definition	Probability	Software Level
Catastrophic	Aircraft destroyed, significant number of fatalities	< 1E-9	A
Hazardous	Damage to aircraft, occupants hurt, some deaths	< 1E-7	B
Major	Pilot busy, interference with ability to cope with emergencies	< 1E-5	C
Minor	Little to no effect on occupants or airplane capabilities	N/A	D
No Effect	No effect on occupants or airplane capabilities	N/A	E

To help achieve a high level of design reliability, the following activities can be followed:

- Planned, organized, consistent way of doing things
- Thought-out process
- Standards

DO-178B also encourages an iterative process between the systems process and the software life cycle. The software and system process should have feedback loops to the other process. The following diagrams highlight the iterative process:

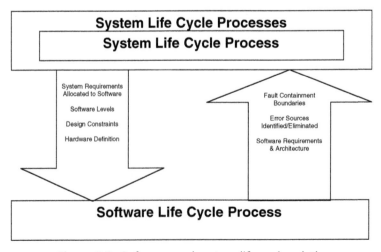

Figure 9-2: Software and system life cycle relation.

System Architectural Considerations

Some architectural strategies may limit the impact of software errors and therefore reduce the software certification level. Reducing the software criticality and the associated software certification level can minimize the required DO-178B process requirements. Various methods are utilized by the aerospace industry to minimize the software criticality level. Various System Architectural methods can be used but as with most things there is a price tag and schedule concern associated with the decisions. Thee determination of the criticality level and associated software certification level has to be done during the system safety assessment process and should consider some of the following methods as options to reduce the software criticality.

Partitioning

This is a technique for providing isolation between software components with independent functionality. If protection by partitioning is provided, the software level criticality may be reduced and each component may have a different software certification level based on the most severe failure condition category associated with that component.

Partitioning in case of safety critical software implies protection; the protection can be done by using different hardware resources such as processors, memory devices, I/O drivers, and timers. The system safety assessment board will have to assess how well the partitioning is done and the kind of protection or un-protection that each partition has.

Multiple Versions of Dissimilar Software

This is a design technique that requires producing two or more versions of the software with the same functionality independently. This will provide redundancy and eliminate having the same errors in the software. Take for example an engine controller with two channels; this is already a redundant system because it has two channels to operate. Once one fails the other will take control of the engine and provide safe operation. The software on both channels can be the same, but it has to be certified to level A because of its criticality. Another option is to have multiple versions of dissimilar software and try to reduce the criticality to lower levels.

Safety Monitoring

Safety monitoring is basically software that is added to protect against specific failure conditions by monitoring a function for failures. The monitoring function can display warnings of potential failures or can switch to redundant functionality if it exists. Monitoring software does not need to be at the same level as the functional software and can be reduced in level as long as it does not cause a system failure. Figure 9-3 shows a basic system that uses monitoring software.

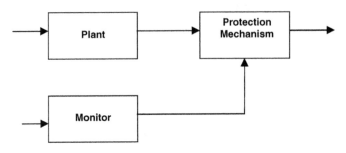

Figure 9-3: Basic monitoring system.

System Architectural Documentation

After an architectural strategy is chosen to limit the impact of software errors and the system safety assessment process establishes a software certification level, the decisions are documented in the Plan for software aspects of certification (PSAC). The PSAC is a document used for initial communication with the certifying authorities on the software project and provides an overview of the program and the software development process. The FAA's Designated Engineering Representative (DER) acts as a liaison between the FAA and the software project. The DER works with the software team when making system architectural decisions and is an active participant in the system safety assessment process.

DO-178B Software life cycle

The DO-178B guidelines impose criteria on the software life-cycle processes to ensure that the development results in a safe system. The processes can be categorized as developmental or integral. The integral processes such as tool qualification, verification, configuration management and quality assurance provide software engineering process support that helps ensure the completeness and quality of the software development activities. The use of off-the-shelf, including commercial off-the-shelf (COTS), software is permitted in the DO-178B framework; however, the software must be verified to provide the verification assurance as defined for the criticality level of the system.

DO-178B enables the guidelines to be tailored to the processes of your organization. The following sections list activities and associated deliverables that are expected as part of the overall DO-178B process guidelines.

DO-178B has different guidelines for the different software certification levels that are determined based on the system safety assessment as highlighted in table 1. For all software levels, some documents and data have to be generated to meet the requirements of DO-178B. Following is a list of the software life cycle data required by DO-178B:

Planning
- Plan for software aspects of certification (PSAC)
- Software development plan

- Software verification plan
- Software configuration management plan
- Software quality assurance plan
- Software requirements standards
- Software design standards
- Software coding standards

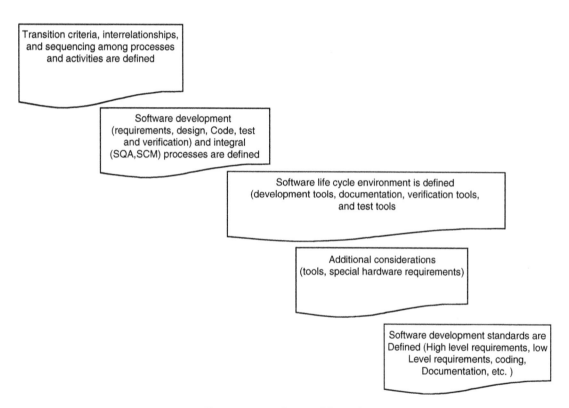

Figure 9-4: Software life cycle.

Development

- Software Requirements (high-level requirements)
- Software Design (low-level requirements)
- Software Design (architecture)
- Software Code

Development process

The software development process consists of the following;

- Software requirements process
- Software design process
- Software coding process
- Integration process

Figure 9-5 shows the allocation and traceability of software requirements.

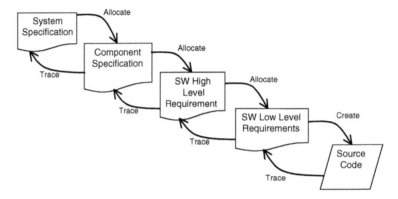

Figure 9-5: Allocation and traceability of software requirements.

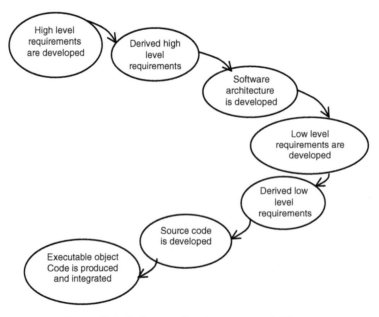

Figure 9-6: Software development activities.

Software Development Activities

Figure 9-6 highlights the software development activities.

Software Requirement Verification

- High-Level Software Requirements comply with System Requirements
- High-Level Software Requirements traceable to System Requirements
- High-Level Software Requirements accurate and consistent
- High-Level Software Requirements compatible with hardware
- High-Level Software Requirements are verifiable

Software Design Verification

- Low-Level Software Requirements comply with High-Level Software Requirements
- Low-Level Software Requirements traceable to High-Level Software Requirements
- Low-Level Software Requirements accurate and consistent
- Low-Level Software Requirements are verifiable
- Software Architecture complies with High-level Software Requirements
- Software Architecture traceable to High-level Software Requirements
- Software Architecture accurate and consistent
- Software Architecture compatible with hardware
- Software Architecture is verifiable
- Software Architecture conforms with Software Standards

Software Code Verification

- Source Code complies with Low-Level Software Requirements
- Source Code traceable to Low-Level Software Requirements
- Source Code complies with Software Architecture
- Source Code is traceable to Software Architecture
- Source Code is accurate and consistent
- Source Code is verifiable
- Source Code conforms with Software Standards

Verification of Integration Process

- Ensure testing methodology of Software Verification Plan was used
- Ensure test cases are accurately developed to meet requirements and coverage
- Ensure verification results are correct

- Object Code complies with High-Level Software Requirements
- Object Code is robust with regard to High-Level requirements
- Object Code complies with Low-Level Software Requirements
- Object Code is robust with regard to Low-Level Requirements
- Object Code is compatible with target hardware
- Object Code complies with Software Structure (where applicable)

Verification of Verification Process

- Test Procedures are correct
- Results are correct
- Compliance with Software Verification Plan
- Test coverage of High-Level requirements
- Test coverage of Low-Level requirements
- Test coverage of object code
- Test coverage—Modified condition/decision
- Test coverage—Decision
- Test coverage—Statement

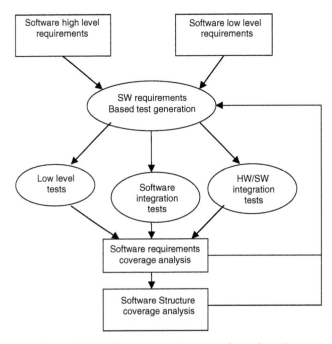

Figure 9-7: Software requirements based testing.

- Test coverage - Data coupling and Control coupling

Configuration Management

- Configuration items are identified

- Baselines and traceability are established

- Problem report, change control, change review, and configuration status accounting are established

- Archive, retrieval, and release are established

- Software load control is established

- Software life-cycle environment control is established

Software Quality Assurance (SQA)

- Assure transition criteria for software requirements process

- Assure transition criteria for software design process

- Assure transition criteria for software code process

- Assure transition criteria for software integration process

- Software quality assurance records

Object Oriented Technology and Safety Critical Software Challenges

Object Oriented Technology (OOT) has many advantages over the traditional software development methodologies. Some of the advantages are ease of use, reuse, working and thinking at the right abstraction level and faster time to market. In addition, the industry is adapting a visual modeling approach for the development of object oriented technology. The visual modeling approach utilizes the use of the Unified Modeling Language (UML). UML is a design notation (language) that allows users to model their design graphically; this adds the advantages of seeing things visually at the right design level. For example, when working with use cases and actors, the user is concerned with specifying the system requirements and should not have any design details at this level. This is very helpful specifically when working with safety critical software and performing traceability between multiple levels of requirements and design.

In addition UML has the benefits of clean portioning between software components. By using a component-based design, the system can be broken into many components that can be developed independently.

The avionics industry specifically and the safety critical application in general are more conservative and careful about adapting new technologies, for obvious reasons. Since the cost is high and the length of safety critical projects is typically long, it is not wise to adopt changes to the existing processes frequently. The stability of the process and tools used to create safety critical software is crucial to the success of the project. On the other hand, OO tools are at a decent maturity level since they have been used for many years.

OO tools can and are being used for a lot of mission critical software and some safety critical software. But OO tools have not been widely adopted in the avionics industry because DO-178B was developed before the widespread use of OOT. I will discuss some of the issues that face OO use in safety critical software, mainly avionics software.

Iterative Process

One of the major advantages of OO is the ability to be used in an iterative and incremental development process instead of a waterfall life cycle. This also promotes the use of component-based architectures, and the use of visual modeling (UML) in software development. An iterative process provides an alternative to the waterfall life cycle where all of the requirements for the system have to be defined up front prior to writing any code. Iterative processes allows users to define requirements, create the design, generate and test the code, and do it all over again with new or changed requirements. It is important to note that the design has to be created based on components and the iteration should be done only once for the specified requirements. There will be integration and testing for the product as a whole but during the creation of the iteration users should avoid redesigning or recoding unless it's absolutely necessary.

The following diagram shows an iterative process using the spiral method:

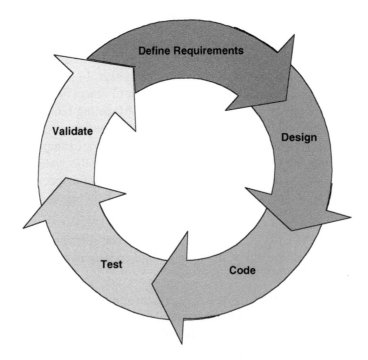

Figure 9-8: Iterative software process.

Issues Facing OO Certification

OO is different from other traditional software development methods, and in a way has many advantages over them. In order for users to develop safety critical software for aviation, they have to perform all of the requirements according to the DO-178B guidelines. These guidelines as highlighted earlier in this paper can be time consuming, expensive and difficult in some instances such as when performing structure coverage at the object level. Another area of difficulty is when performing traceability between the object code and source code required for safety criticality level software. Since the introduction of OO or UML all the users now work at a higher abstraction level and some projects use tools so that the code and test can be generated automatically. In these instances when projects are utilizing DO-178B, there are a few issues that the user needs to be aware of. Following is some discussion on the various issues related to using OO in conjunction with DO-178B.

Auto Code Generation

Auto code generation is not new to software development, and many companies have previously used auto code generation in safety critical software. Examples of systems that have helped the avionics industry auto generate code are Software through Pictures and other UML-based tools. Most of these companies have spent millions of dollars on creating the auto code generation tool framework and the associated DO-178B required certification package.

When using UML or OO there are many advantages to generating the code automatically, and there are a few tools on the market that provide partial or complete code generation. This is possible because UML is a standard formal language and supports Statecharts which are executable. Some of the advantages of using auto code generation are:

1. Keeping the code and models linked. This is important because in most instances when the code is done manually the design is left behind and not updated. In a way this is a difficult and redundant issue because the developer has to first make a modification to the design and then make the same change to the code. For example, to add a new function to a UML diagram, the user would add it to the diagram first, then add it in the code. When using an auto code generation toolset, the user would add it in the design and the code would automatically be generated correctly for the updated design. If the user is using Statecharts then the behavior of the function can be defined. If the user is using a tool that generates code from Statecharts then the code will also be automatically generated from the updated Statecharts.

2. Improve traceability, since the code is generated directly from the design which makes it easy to maintain the traceability between the design and code.

3. Auto code generation will improve productivity by eliminating a manual step from the software development process.

When using auto code generators, there are few things a project manager should watch out for:

- Does the code generator meet the guidelines of DO-178B or the guidelines for your application?

- Since auto code generators have the potential of generating the wrong code and inserting errors into your application, they have to be certified to the same level as your application per the requirements of DO-178B for tool qualification.

- If the project is using a partial code generator that generates the frames only from the UML or OO diagrams, a user should make sure that the reverse engineering capability is tested. In this case a user would generate the code frames and add implementation details. Once the user modifies the design, the user will have to generate the frames again. The tool should not disturb the implementation and should only generate the changes.

- If you are using a complete code generator then you have a few things to worry about:

    - Certification per the DO-178B requirements for tool qualification

    - Can the user do everything needed using only the code generator?

    - If the user modifies the code, can it be reverse engineered to the model?

    - Does the user have control over every configuration item? This is important from a maintainability point of view. After the user would want changes to have a minimal impact on the rest of the design and code.

These are only some of the things that need to be considered, but the list is too exhaustive to discuss here. There are additional issues when using auto code generation tools concerning dead/deactivated code and requirements based testing.

Auto Test Generation

Some of the OO or UML based tools can also auto generate test from class diagrams, Statecharts, activity diagrams, and other diagrams. It is important to understand that generated code from the design is not considered to lead directly to requirements-based testing. Generating the test from the use cases if they are written correctly may be acceptable. For safety critical software it is important to have clear testable requirements and then to be able to base the testing on these requirements and show clear traceability.

Traceability

Most of the UML-based tools or visual modeling tools don't provide the mechanism to tag and perform traceability to requirements. This may become a problem if you don't find an automated way to perform the traceability and keep it up to date with your UML diagrams. One way to solve this problem is to export the software requirements including derived requirements to a tool that is capable of performing and maintaining the traceability. The trick here is to keep the two activities connected where changing one will change the other, this way the traceability will always be up to date.

Not everything in your UML diagrams will have to be exported; you also need to pay special attention to the abstraction levels of your diagrams. The source code will have to be traced to the low level design (functions within classes) and then the high level design (class diagrams or packages) will have to be traced to the requirements which may be documented using use cases or text.

Configuration Management

When using visual modeling tools the user will have to control both the models and the generated source code in addition to any test cases and documentation. Projects should have a configuration management environment where the user can save all of the generated artifacts in one place.

Structural Coverage

Structural coverage is important because it can show which code has not been tested and which code has not been accessed or tested. The user will either have to write more tests or deal with the non-executed code as dead or deactivated code. Structural coverage in DO-178B depends on the level of certification. For Safety Critical Level A software, the structural coverage has to be done at the modified condition decision coverage and the object code has to be traceable to the source code.

Dead/Deactivated Code

Dead and deactivated codes are found after a complete requirements-based testing is pre-formed. All of the code that is not executed during the testing will be classified as dead code and has to be removed unless it can be classified as deactivated, which means that it is in the software for a purpose. Deactivated code needs to be activated at some time and utilized. For example, the growth of an application may contain code for future models. Deactivated code will have to have requirements and be captured in the design. The deactivated code must be tested to the associated requirements.

When using OO languages such as C++, the compiler and linker may insert code into your object code to support creation of functionality such as dynamic dispatching. DO-178B does not require certification of compilers because the output is tested according to the requirements of the system. In the case of OO the compiler may be inserting code that cannot be reached or tested and will be considered dead or deactivated code depending on the situation.

Inheritance and Multiple Inheritance

Single inheritance should not be an issue when designing safety critical software and may have some advantages. For example, inheriting an interface to provide an implementation is perfectly expectable. The problem comes when using multiple inheritances where a single class can inherit from two or more classes. This may cause problems such as having the same function signature in both classes which may case problems and ambiguity. Avoiding multiple inheritances should keep the design clean and should make the traceability and requirements based testing straight forward.

Summary

Although this article has only touched the surface of the safety critical software development process following the DO-178B guideline, we have discussed the high level points to watch for during process development. The DO-178B guideline is very comprehensive and provides a framework for developing safe code. A few draft position papers exist today on the subject of using OOT in safety critical software. There is a lot of work to be done on defining the extra care needed when using such technology in aerospace or other mission critical software. If software requires certification by the FAA, the project team needs to plan ahead and communicate as much as possible through their DER (designated engineering representative) to the certification authorities. The FAA required PSAC for certification should address all of the potential issues of certification so that they can be resolved and agreed to at the beginning of the project.

Another requirement for a successful project is to have solid requirements, design, and coding standards. The coding standard should meet the industry-accepted standard for safety critical software including what the industry considers a safe subset of the programming language. Some good references for coding standards currently exist, such as the one published by David Binkley of the National Institute of Standards and Technology (NIST) titled "C++ in Safety Critical Systems." Design guidelines should define a subset of UML and the project will use UML. For example, projects should limit the number of inheritance and the number of classes in a package since this will help you perform the traceability, testing, and structure coverage required by DO-178B.

Overall, OO is a good choice for safety critical systems if used correctly and wisely, but thought must go into developing a software development plan that meets the requirements of the DO-178B guidelines.

References

1. RTCA/EUROCAE: DO-178B *Software Considerations in Airborne Systems and Equipment Certification.*

Installing and Using a Version Control System

By Chris Keydel and Olaf Meding
info@esacademy.com
Embedded Systems Academy
www.esacademy.com

Introduction

Software configuration management (SCM or CM) is the discipline of managing and controlling the evolution of software (over time). It is many things to many (different) people and encompasses process, policy, procedure, and tools. CM is a cornerstone in software development, and thus has a very broad spectrum. The focus here is the embedded systems market and the software developer rather than a management perspective. Advice and concepts are more practical rather than theoretical.

Three central and key concepts of CM are versioning, tracking, and differencing. Versioning is the idea that every file or project has a revision number at any given state in the development process. Tracking is essential because it provides a central database to track and manage bug reports, enhancement requests, ideas, and user or customer feedback. Tracking is needed to maintain current and past releases and to improve future versions of a software product. Finally, differencing allows the developer to easily view the difference between any two revisions of a file or project. Many of the more advanced CM concepts integrate, manage and track these core development activities.

Improving the quality of your code and boosting developer productivity are not mutually exclusive goals. In fact, the opposite is true. Both quality and productivity dramatically increase with the knowledge and the applications of the tools and concepts outlined below. The right tool will be fun to use while at the same time increase the engineer's confidence that a software project will achieve its goals. This paper will show how basic and proven tools and procedures can be used to produce high quality software. From my own experience and by talking to colleagues I noticed that many CM tools, twenty years after their introduction, are still not being used to their full capabilities. This seems to be especially true in the Embedded Systems and PC (non-Unix) world. A few of the reasons are: small (often single person) teams or projects, a wide variety of platforms and operating systems, budget constraints, including the cost of commercial tools, non supportive IS departments, prior negative experiences with hard to use CM tools, and lack of support from management. A trend in the development of software for embedded systems is the increased use of personal computers (PCs). More and more software development tools for PCs are becoming available and are being used. This is good news in that excellent and intuitive CM tools are also now available. This paper seeks to outline such tool categories, including how they are used, along with some simple commonsense procedures. It is probably safe to assume that most source code including firmware for embedded systems and microcontrollers is edited and managed on PCs these days. Therefore, all concepts, ideas, and tools discussed below apply as well to these environments.

This paper is divided into sections outlining version control overview, version control client issues, version control concepts explained, tips, bug tracking, non-configuration management tools, and finally a reference section. Other software development tools will be briefly mentioned and explained precisely because they too increase quality and productivity, but are usually not considered CM tools. The lack of their use or even knowledge of their existence, among many fellow professional software developers, motivated the inclusion. These tools include mirror software, backup software, World Wide Web browsers, and Internet News Groups.

The Power and Elegance of Simplicity

Before discussing version control (VC) systems in detail, it is worth noting that standard VC tools can be (and are) applied to all kinds of (text and binary) files and documents, including World Wide Web pages. The major and outstanding advantage of standard source code over any other kind of document is the fact that it is strictly and simply (ASCII) text based. This unique "feature" has two major advantages over other kinds of documents (HMTL included), differencing and merging (explained in detail later). Differencing allows non-proprietary tools to show the difference between any two files. Secondly and equally important is that standard non-proprietary or generic tools can be used for merging a branch file back into the main trunk of development (also defined below). An unfortunate but minor exception is the line ending conventions used by prevailing operating systems, CRLF in PCs, CR in Macintosh, and LF in Unix. Finally, many version control systems take advantage of the simple structure of standard text files by only storing deltas between file revisions and thus significantly reduce disk space requirements.

Version Control

Central to configuration management is the version control (VC) tool. VC captures and preserves the history of changes as the software (source code) evolves. VC empowers the user to recreate any past revision of any file or project with ease, 100% reliability, and consistency.

A common misconception is that VC is not needed for projects with only a single developer. Single developers benefit just as much as teams do. All the reasons for using VC also apply to single developers. Many VC systems are easy to setup and use and the VC learning curve is significantly smaller then any single side effect (see below) of not using a VC system.

Finally, how safe is the assumption that a single developer will maintain a certain software project throughout its entire life cycle? And even if this holds true, how likely is it that this developer will be available and able to remember everything in an emergency case? Corporate productivity and software quality are issues as well.

Typical Symptoms of Not (Fully) Utilizing a Version Control System

Here are some telltale symptoms that indicate only partial or perhaps wrong use of VC or the complete lack of a VC tool. These symptoms include manual copying of files, use of revision numbers as part of file or directory names, not being able to determine with 100% confidence what code and make files were used to compile a particular software release, and bugs considered fixed are still being reported by customers. Also, checking in source code less frequently than multiple times per week and developers creating their own sub-directories for "testing" indicate serious problems with the (use of the) VC tool. Two other symptoms include the inability to determine when (date and release) a particular feature was introduced, and not being able to tell who changed or wrote a particular (any) section of code.

Simple Version Control Systems

Basic and single-developer VC systems have been around for a long time. One of the simplest, oldest, and most successful VC systems is the Revision Control System (RCS). RCS is included with most Unix distributions and is available for many, if not all, other operating systems. Many modern and full-featured VC systems are based on RCS ideas and its archive files. The key idea is simple. Every source file has a corresponding archive file. Both the source file and its archive file reside in the same directory. Each time a file is checked in (see below) it is stored in its corresponding archive file (also explained below). RCS keeps track of file revisions (AKA versions) and usually includes tools for file differencing and file merging. However, the early RCS systems were not well suited for teams, because of the location of archive files and the inability to share them between developers. Also, there was no concept of projects, i.e. sets of related files, representing a software project. Good documentation is readily available and most implementations are free and easily available and can be downloaded via Internet FTP sites.

Advanced Version Control Systems

Advanced VC systems offer wide variety of options and features. They fulfill the needs of large teams with many (thousands of) projects and files. The team members may be geographically distributed. Advanced VC systems support a choice of connectivity options for remote users and also internally store time in GMT or UTC format to support software developers across time zones. Security and audit features are included, so access can be limited to sensitive files and to provide audit trails. Promotion models are used to manage software in stages, from prototyping to development to final testing, for example. Parallel development models allow all developers to work on independent branches and full featured and automated merge tools ease the check-in process. Some VC systems employ virtual files systems that transparently point or re-direct the editor or compiler directly into a SQL strength database. Another useful feature is the ability to rename or move files within the version controlled directory tree without losing the integrity of older revisions of a file that was moved or renamed. Powerful build tools are also frequently included to help automate the build (compile) process of large software projects. Event triggers alert the software team or a manager of key events, such as a check-in or a successful nightly build. It is not uncommon for large projects to employ a full-time dedicated configuration administrator.

What Files to Put Under Version Control

This is a subject of many debates. Many published CM papers include a full discussion of this topic, so there is no reason to expand on this topic here. In brief summary, agreement exists that all source code files needed to create the final executable must be added to the VC system. This includes, of course, make files. A good VC system should make it very easy to re-create the final executable file, so there should be no need to put binary intermediate files (object files, pre-compiled header files, etc.) or even the final executable file itself under VC. Library files, device drivers, source code documentation, any kind of written procedure, test scripts, test results, coding standard or style guides, special tools, etc. can and probably should be added to the VC system.

Sharing of Files and the Version Control Client

Most VC systems are true client/server applications. Obviously, in a team environment, all developers somehow must have both read and write or modify access to the file archive located on a central file server. So, you might ask yourself exactly how files are transferred from that central file server to your local PC? The VC system vendor often provides a VC client that provides a user interface to exchange source code between the PC (workstation) and the repository (explained below) on the central file server. However, this VC client, may or may not be accessible from your PC and its operating system. And the PC may or may not have direct access to the server's file system (NFS in Unix or drive mappings with Microsoft LANs and Novel's Netware). Physical location of both the client PC and the file server affect and also limit the choice of available options. In summary, there are two variables—whether the VC client can run on your local PC and whether your PC can access the file server's file system. The two variables combine into the four choices or variations outlined below.

No Local Client and No Common File System

This is the least desirable scenario. An example would be a PC accessing an older mainframe. Or a much more likely situation is a geographically separated team with a limited budget. Here the only choice is to use Telnet or some kind of remote Unix shell to run the client directly on the central file server or on a local workstation (with respect to the file server) and then manually FTP the files between the file server and your PC. For example, login to the file server, checkout a file, FTP it to your PC, change it, FTP it back, and then finally perform a check in.

No Local Client But Common File System

This is similar to the first scenario. However, the PC can access files directly. The applications on the PC (editor, compiler, etc) can either access source files across the LAN or the user has to manually copy them. The first method is slow and will increase network traffic. The latter requires extra and manual steps after checkout and before check-in.

Local Client But No Common File System

The problem here is that few VC systems provide clients capable of accessing the server over a LAN or across the Internet, via a network protocol like TCP/IP. Fewer VC systems still use secure methods of communications (public keys instead of un-encrypted passwords during login and encrypting file content instead of transferring text).

Local client and common file system

This is the most common and flexible scenario. The client often takes advantage of the common file system to access the repository on the file server. The client may not even be aware that the repository is on a separate file server. The VC client is much easier to implement by a VC provider, which in part may explain its popularity.

Integrated Development Environment Issues

Integrated development environments (IDEs) for software development predate graphical (non-character-based) windows and became very popular during this decade, the decade of the graphical user interface (GUI). Most IDEs provide an excellent and integrated source code editor with many hip features, such as automatically completing source code constructs while editing and tool tip type syntax help displayed as you type. In addition, they usually feature well-integrated debuggers, full and finely tunable compiler and linker control, and all kinds of easily accessible online help. What is often missing in such IDEs, or was added only as an afterthought, is an interface to a VC system. The Source Code Control (SCC) specification (see below) is one such integration attempt. Another major problem with many modern IDEs is proprietary, in some cases even binary, make file formats. These proprietary make files are much more difficult to put under version control. Problems with these make files include merging, sharing when strict locking (see below) is enforced, and the fact that each IDE typically has a different and proprietary format. In fact, make file formats often change between each release of an IDE from the same vendor. IDEs are a nice and convenient tool and there is no doubt that they increase developer productivity. However, they often produce a handful of proprietary files, containing make information, IDE configuration information, debug and source code browse information, etc. And it is not always obvious exactly what information these files contain and whether they need to be put under VC.

Graphical User Interface (GUI) Issues

A number of VC vendors provide GUI front-end VC clients. These graphical (versus text-based) clients are much more difficult to implement well and are therefore often expensive. They can easily exceed the budget of an embedded systems project. They occasionally appear clumsy and sometimes only support a subset of features compared to an equivalent text based (command line) VC system. Another problem is that they may not be available across operating systems and hardware platforms.

Common Source Code Control Specification

The Microsoft Common Source Code Control (SCC) application-programming interface (API) defines an interface between a software development IDE and a VC system. Many Microsoft Windows development environments (IDEs) interface to a VC system through the SCC specification. The VC system acts as a SCC service provider to a SCC aware IDE. This enables the software developer to access basic VC system functions, such as checkout and checkin, directly through the host environment. Once a SCC service provider is installed, the user interface of the IDE changes. Menus are added to the IDE for the operations offered by the SCC service provider. And file properties are extended to reflect the status of the files under source code control.

The problem with this approach is that often only a subset of VC functions are provided through the SCC interface. Therefore the standalone VC client is still needed for access to the more advanced functions offered by most VC systems. So why not always use the standalone VC client with access to all features?

The VCC specification is only available to Microsoft Windows users since this a Microsoft specification. Furthermore, the documentation for the SCC API is not publicly available. To receive a copy, you must sign a nondisclosure agreement. It should be noted that this is not the only way to interface IDEs to VC systems. Some VC vendors call their service interface extensions, because the VC system extends its functionality to an IDE or a standard file browser offered by many operating systems.

World Wide Web Browser Interface or Java Version Control Client

This sounds like a great idea. In theory, well-written Java code should run anywhere either as a standalone application, or in a web browser as an applet, or as a so called plug-in. The VC server could be implemented as a dedicated web server, which manages the VC archive, repository or database (explained below). Another advantage of Java is build-in security. However, for various reasons there are very few Java applications of any kind available that achieve the goal of write once and run anywhere. A select few VC vendors do provide Java based VC systems, but their user interfaces appear sluggish and somewhat clumsy. It should be noted that web browsers rarely support the latest version of Java, so the plug-in must be written in an older version of Java. In summary, Java offers many attractive features. I would expect to see more Java based VC systems in the future as the Java and related technologies, such as Jini, mature.

Version Control Concepts Explained

Revision, Version, and Release

All three terms refer to the same idea: namely that of a file or project evolving over time by modifying it, and therefore creating different revisions, versions and eventually releases. Revision and version are used interchangeably; release refers to the code shipped (or released) to customers.

Checkout, Check-in, and Commit

Checkout and check-in are the most basic and at the same time the most frequently used procedures. A checkout copies the latest (if not otherwise specified) version of a file from the archive to a local hard disk. The check-in operation adds a new file revision to that file's archive. A check-in is sometimes called a commit. The new version becomes by default, if not otherwise specified, either the head revision if you are working on the trunk, or a tip revision, if you are working on a branch (all explained below). When a file is checked into a revision, other then the head or tip revision, a new branch is created.

Lock, Unlock, Freeze, and Thaw

The lock feature gives a developer exclusive write (modify) access to a specific revision of a file. Some VC systems require (or strongly recommend) and enforce a strict locking policy. The advantage of strict locking is prevention of file merges and potential conflicts. These VC systems often also provide an administration module that allows for enforcing and fine-tuning of policies. An example of such a policy is allowing only experienced developers to make changes to sensitive files. This approach works well in smaller teams with good communication. Here, the team (management) decides beforehand who will be working on which file and when. Each team member knows or can easily tell who is working on what file. One major drawback is that if someone else already locked the file he needs, a developer must ask and then wait for another developer to finish and check in her changes. Frozen files cannot be locked and therefore cannot be modified until thawed. This may be a useful management tool to let developers know that certain files should not be changed, say right before a major release.

Differencing

To be able to tell the difference between two revisions of a file is of most importance to any software developer, even if for whatever reason no VC system at all is used. In general, the difference tool allows a developer to determine the difference between two files and, more specifically, between two revisions of the same file. Any VC system has a built in difference tool. However, there are many difference tools available. Some tools are text based and others use a graphical (visual) user interface. It is a good idea to have a selection of different difference tools handy, depending on the task to be performed. The build-in tool is often sufficient and convenient to access through a menu choice for quick checks. More powerful difference tools can compare entire directory trees and may format and present the output in a more readable form.

Difference tools help avoid embarrassment by pointing out the debug or prototype code you forgot to remove. Often there is an overlap between developing a new feature and debugging a problem recently discovered or reported. For example, if someone asks you to find a high profile bug while you are working on implementing a new feature. Using a difference tool before each check-in helps prevent forgetting to take out code not meant for an eventual release to customers, such as printing out the states of (debug) variables at run time.

Another important use for a difference tool is detecting all code changes between two revisions or entire set of files that introduced a new bug. For example, customer support enters a defect report into the bug tracking database (see below) stating that a bug exists in release 1.1 but the same problem does not exist in release 1.0. The difference tool will show all sections of code changes that most likely introduced the bug. Being able to quickly locate a bug is important for Embedded Systems developers, where hardware is often in flux, because it may not be initially obvious if the bug is caused by a software change, hardware change, or a combination of the two.

Build and Make

A powerful build tool plays a vital part in many CM setups and more advanced VC systems include flexible and powerful build tools. A make tool is standard on Unix and also shipped with most compilers. However the standard make tools are often less capable compared to full-featured build tools included with VC systems. Both build and make tools require some careful planing, but once correctly implemented provide the ability to build with ease any release of an application at any time. The goal is to ensure consistent and repeatable program builds. Another advantage is automation of large-scale projects. Many software manufacturers perform builds of entire software trees over night. The results of these nightly builds are then ready for examination on the following workday.

Trunk, Branch and Merge

Merging refers to the process of consolidating changes made to the same file by either different developers or at different times, i.e. revisions of a file. A merge is typically required to copy a change made on a branch back to the main trunk of development. Usually each change is in a different area of the file in question and the merging tool will be able to automatically consolidate all changes. Conflicts will arise for overlapping changes, however, and they must be consolidated manually.

There are three main reasons for branching. They are changing the past, protecting the main trunk from intermediate check-ins, and true simultaneous parallel development. Most, if not all, VC systems support branching as a way to change the past, that is to create a branch to add bug fixes (but not necessarily features) to an already released software product. For example, a bug fix results in release 1.1, while the main development effort continues on release 2.0. A merge may be required if the bug was first fixed for release 1.1. In this case, a developer would either manually copy or better use a merge operation to copy the code representing the bug fix into release 2.0.

Some teams decide that all new development should take place on a branch to protect the main trunk from intermediate check-ins. This approach requires merging code from the branch to the main trunk of development after completing each specific feature (set). Often teams decide to check-in all code to their respective branch at the end of every workday to protect themselves against data loss. A major advantage of this approach is that the trunk contains only finished changes that compile correctly and that are ready to be shared within the team. This approach also rewards frequent merging, because the longer a developer delays a merge, the higher is the chance of encountering merge conflicts.

Not all VC systems support true simultaneous parallel development, however, and there is some controversy and reluctance by some programmers to use or take advantage of branching, because of potential subsequent merge conflicts. Notably, VC systems requiring exclusive write access (see strict locking above) before a change can be made do not support parallel development. On the other side, VC systems that do encourage simultaneous and independent development make extensive use of branching. This can result in minor conflicts while merging changes made on a branch back to the main trunk of development. However, minor conflicts are a small price to pay for large teams in exchange for the ability to work in parallel and independently from each other. In general, branching depends on frequent and thorough developer communications and it is a good idea to implement a consistent naming convention for labeling (see below) branch revisions.

Files and Projects

Some VC systems strictly handle only individual files. At best, all related files are located in a single directory (tree). This is often sufficient for smaller projects. Large projects, however, are often organized across a wide and deep directory tree and often consist of sub-projects. Therefore, a VC project is defined as the set of files belonging to the corresponding source code project. And a source code project is defined as the set of files that are compiled into the target executable, i.e. device driver, dynamic link library (DLL), static library, program, etc. More advanced VC systems support projects or modules, so that there is an one to one relationship between a source code project and the corresponding VC project. This greatly simplifies VC operations such as check-in, checkout, and tagging (see below). For example, instead of tagging a specified number of individual files and directories belonging to a source code project, the user can just tag the corresponding VC project in a single transaction.

Checkpointing, Snapshots, Milestones, Tagging, and Labeling

The goal here is to be able to retrieve the state (certain revision number) of all files of a project in the future. This is a very important activity because it allows a developer to retrieve all files that comprise, for example, release 1.0 of a project. Or a test engineer could mark all files that were present and their revision number when a certain defect was found. Not all VC systems fully support this critical feature. There are two common implementations. The first is to record all revision numbers of all files involved. A drawback of this method is that the checkpoint, snapshot, or milestone information has to be stored somewhere in a special VC file or database. The second implementation of the critical feature is to mark all revision numbers of all files involved with a certain label or tag. A VC system that favors this method has to be able to store many thousands of labels for each revision of each file. The label information is stored inside the archive file. Though all four terms refer to the same idea, they are often used interchangeably for both methods. Note that some VC systems allow a particular label to be used only once in a file's history and some VC systems limit any one revision to a single label.

Archive, Repository, and Database

The archive is a file containing the most recent version of a source code file and a record of all changes made to it since that file was put under version control. Any previous version of a file can be reconstructed from the record of changes stored in the archive file. Sometimes the file tree containing the archive files is called the repository. Some VC systems use a database, instead of individual archive files, for recording the history of source code changes. With the possible exception of single user systems, the archive is most frequently stored on a central file server. Usually, all that is needed to re-create any particular version of a file is that file's archive. Backup of archive data is therefore vitally important. A network administrator usually performs a daily backup as part of the file server maintenance. This frees the software engineer from this time consuming task. Some VC systems specifically designed to support physically distributed teams use archive replication as a way to share archives. In this scenario, replication software resynchronizes the archive every few hours between the central file server and a remote location file server. The advantage of this approach is faster access for the remote team of developers. Drawbacks include the complexities of installing and maintaining the replication software and potential re-synchronization conflicts between file servers.

Working Copy and Sandbox

The working copy contains a copy of each file in a project at a specified revision number, most commonly the latest revision number of the main trunk or of a branch. Working copies provide a way for each developer to work independently from other developers in a private workspace. The working copy is usually located on the developers local hard disk and is created there through a checkout operation. A working copy is sometimes also called a sandbox. It is not uncommon to have multiple working copies per developer per project, with each working copy at a different trunk or branch release. I often create a new working copy immediately after checkpointing a major project release to double check and verify all file revisions. All files in the new working copy should be identical to all files the working copy that created the new release (see tips below).

Event Triggers

Many VC systems provide a means to execute one or more statements whenever a predefined event occurs. These events include all VC operations and more. For example, emailing a project manager each time a file is checked in.

Keywords

A keyword is a special variable that can be included within source code. This variable is embedded usually between two $ characters (for example $Author$) to represent textual information in a working copy. Keywords can optionally be expanded, i.e. replaced with their literal value, when a revision is checked or viewed. This is a very useful feature, because it allows the VC system to place information directly into the source code. At the same time keywords allows the developer to select what VC information, i.e. which keyword(s) should be inserted and where it should be located in the source code, usually at the beginning or at the end of a file.

Tips

Source Code File Directory Tree Structure

One key aspect of any larger software development project is the implementation of a directory structure and layout. This directory tree reflects the relationship of the source code files that define a software project. Most importantly, there should be no hard coded paths for include file statements in C/C++, import file statements in Java, or unit file statements in Pascal source code. Instead, a relative path statement should be used so that (multiple) working copies of files and projects can be created anywhere on a software developer's local hard disk. Placing shared source code files in the directory tree is an art, because each method has advantages and disadvantages and much depends on the specifics of a project's architecture. Therefore, it is not uncommon to adjust the source code file directory tree as the project grows. However, not all VC systems support re-naming and moving of source code files well. An article by Aspi Havewala, called "The Version Control Process" (see references below) illustrates one nice way to organize shared source code and C/C++ style header files.

More Applications of the Difference Tool

For extra safety and protection of source code and its long term storage, I sometimes backup my archives on the central file server and my working files on my PC to a writeable CD. CDs are nice because they have a much larger shelf life then any other backup medium. However, for a number of reasons CD burners sometimes fail to make exact copies. This is where a tool such as Microsoft's WinDiff (see resources below) or similar tools provide assurance and confidence in backups needed to remove obsolete files from your hard disk. WinDiff will read (and in the process verify) every single byte on the new CD in order to confirm that all files were copied correctly. Being able to remove obsolete code efficiently frees time for the developer and therefore increases productivity and also helps the software engineer to focus on the main project.

Another useful application of WinDiff is verifying what files were copied where during installation of new software (tools) to a critical PC. This not only provides extra insight on how the new tool works, it might also be a great help to debug incompatibility problems between various applications, which has been and continues to be a major problem for Microsoft operating systems. For example, installing a new compiler or other software development tool may consist of many files being copied into an operating system specific directory. To find out exactly what files were installed where, run a recursive directory listing and re-direct the output into a file. For example, type "dir /s/a c:\ > before.dir", install the new tool, run another directory listing, but this time redirecting the output into a file called after.dir. Then use your favorite differencing tool or WinDiff to display the delta between before.dir and after.dir. You can use the same procedure to debug and verify your own installation procedure before you ship your product to a customer.

Heuristics and Visual Difference Tools

It is worth noting that GUI-based difference tools rely on heuristics rather than on exact algorithms to detect and display code that moved from one location to another. Heuristics are also used to differentiate between changes and additions for displaying each in a different color. For example, consider these two revisions of the same file:

Revision 1.1	**Revision 1.2**
line 1 code	line 1 code
line 2 code	line 2 code changed
line 3 code	a new line of code inserted here
line 4 code	line 3 code changed
line 5 code	line 4 code
	another new line of code
	line 5 code

A visual difference tool will have no problem detecting the new code inserted after line 4. However, how can the tool tell the difference between new code inserted and code changed between lines 2 and 3? The new line could either be part of the line 2 change or the line 3 change, and it could also be new code inserted! There is more than one answer here and that is why a heuristic is used to display change codes and colors. So watch out when using a visual difference tool when searching for code additions displayed in, say, a blue color. There may be unexpected surprises; in the above example added code may be shown as changed code. Try the above example or variations of it with your favorite visual difference tool.

No Differencing Tool

On more than one occasion, for example when visiting a customer, I needed to compare two versions of a section of code, and the only tool available was a printer. No problem. I just printed out both sections and then held both copies overlapped against a bright light source. By just moving one of the pages slightly back and forth over the other I was able to easily identify the one line of code that changed. This will also work with a computer monitor displaying two revisions of the same file each in its own editor window. All you have to do is display the code in question in two editor windows one exactly on top of the other and then rotate quickly back and forth between the two windows. Your eyes will easily be able to detect the differing code (if any) as you cycle back and forth between the two windows.

Complete Working Copies

It is a good idea to include any file—i.e., libraries, device drivers, configuration (.ini) files, etc.—that are needed to compile, test, and run a project into the VC system. This ensures that a working copy once created by the VC system is immediately ready for use.

Multiple Working Copies

It is not uncommon to keep multiple working copies of a project on the local PC. Each working copy contains a different release of the project. This simplifies the creation of a list of all changes between a new release and a previous release if you have a working copy for each. The difference tool will point out all sections of code that changed between the two releases. This makes it much easier to document all changes for the new release. Multiple working copies are also very handy for locating bugs in prior releases. Furthermore, I often create a new working copy right after creating a new project release. I then compare (using the difference tool) the new release working copy with the working copy the created the new release. These two working copies better be 100% identical. On reason that they may not be exactly alike is adding a new source code file to your project but forgetting to put that new file under version control.

Style Guide and Coding Convention

Both terms refer to the same document. One of the more important CM documents spelling out coding conventions and styles as a common language, so that all developers working in the same company or on the same project understand each other's source code. This document helps to keep all code in a consistent format and, therefore, increases the overall quality of source code. Many papers and books have been written on this subject and the important thing to remember is that each software project should have a coding convention. This document should be put under VC as it evolves. It is worth noting that source code should be written with humans in mind, not machines. And a style guide helps to achieve this goal.

Bug Tracking

Bug tracking (BT) encompasses the entire project timeline—from the past to the present and into the future. Being able to track an issue into the past enables customer service to determine if, for example, a certain bug is present in the version currently being owned by a customer. Looking at the present tells a manager what issues are currently open and who is eliminating what problem or implementing (working on) that hot new killer feature. And the element of future is represented by feature requests (wish list) or product ideas, both of which are forming the base for future releases, which in turn provide the revenue needed to maintain a software project.

BT is also known as defect tracking. Both terms are somewhat misleading because they imply problems. And yes, problem tracking is vital. However, just as important and worth tracking are enhancement requests, customer feedback, ideas developers have during development or testing, test results, sales force observations, etc. In short, any kind of issue that documents the behavior of past versions and that might help to improve a future product release. Tracking these issues is mainly a database exercise with emphasis on powerful database query capabilities. A BT application provides a central database to track the issues mentioned above. Ideally, the database must be easily accessible to any person involved in a software project, i.e. developers, testers, managers, customer service, and sales personnel. And yes, the user and customer should also have (limited) access to this database.

Significant recent improvements have been made to database (server) technology, local and wide area networks, and (thin) clients capable of running on any desktop PC. So, the technology is available (and has been available for years), yet, BT software lags behind VC software and only a few packages offer tight or complete integration between the two. An example of good integration would be a single piece of client software that can access both the VC archive and the BT database. Another example would be the ability to easily locate the code that closed a certain bug report. Part of the reason for this might be that BT is not as clearly defined as VC. Effective BT requires a real client/server type database (versus a special purpose yet much simpler archive or repository typically used by VC). Another problem might be the wide variety of team requirements of this database, such as what fields a bug report should have, (date, time, who reported a bug, status, etc.), management type reporting capability, ease of constructing queries, database server hardware requirements and database administrator labor.

Equally important is the ability to determine if a specific defect (characteristic) or feature is present in a specific version of a product. This way customer service will be able to respond quickly to specific questions.

I have not personally worked very much with BT tools and my projects and productivity have suffered greatly. The fact that BT tools require some effort to set up and configure for your environment is no excuse for not using this essential tool. I am currently working on this problem and plan to present my findings in a future paper or presentation.

Non-Configuration Management Tools

Mirror Software

Most editors support renaming the current file with a .bak (for backup) extension before making a change. This, at a minimum, allows you to revert back to the last saved copy of a file in case you lost control or your PC crashes while editing source code. Taking this idea one step further, there are source code editors (for example, the Borland BC5 IDE) that automatically save your current file in two places. This feature makes it possible to recover from both a hardware failure and software crash. I fail to understand why this simple and easy to implement feature (saving the current file to a second configurable place, preferably a different hard drive) is not available in all source code editors. However, there are standalone third-party products that automatically mirror any file that changed. One such product is as AutoSave by V Communications (see resources below).

Automated Backups

A software developer should not have to worry about backups of his or her daily work. Having to back up your work to floppy disks or some other removable media is counter productive and cause for many problems. Central backup of software archives, repositories, and databases is a must-have, and there are many applications available for this. If there is no regular backup of this critical data, make sure you copy your resume file to a floppy disk. You may need it soon.

World Wide Web Browser

World Wide Web browsers are a must-have tool to access all kinds of freely available (product support) information, including many forms of technical support. This tool is included here for obvious reasons and for completeness.

Internet News Groups

Internet news groups date back to the earliest days of the Internet. These news groups—a more descriptive name would be discussion groups—use the network news transfer protocol (NNTP) to exchange messages. There are now two types of news servers, vendor specific and traditional. The vendor specific news servers are stand-alone servers with one or more news groups dedicated to the products offered by that company. All you need to access these news servers is the address of the server, usually advertised on the product support web page (see resources below). Once connected it is straightforward to find the news group you need, because there are typically only tens of them with obvious names. Traditional news servers, on the other hand, are all interconnected with each other to form a world wide network. They often contain tens of thousands of news groups, and are maintained by your Internet Service Provider (ISP). A typical news group is comp.lang.c for the "C" programming language. There are also moderated (censored) news groups that filter out all spam (Internet slang for unwanted advertisements) and other non-related offensive message posts. News groups are an excellent (and free!) source of help to any kind of question or problem you might have. You often find an answer to your a.m. posting in the afternoon and messages you post before you leave work are frequently answered over night. You will discover that answers to your questions come for all corners of the world (Russia, Germany, New Zealand, and India, to name a few). Often, while browsing newsgroups, you see posted questions that you know the answer to and in turn help someone else out there to improve her code. News readers are available standalone and are also part of most World Wide Web browsers. Chances are that one is already installed on your PC and waiting for you to use it. Embedded systems software developers are often isolated or work in small teams. News readers open the door to the world in terms of help and support, and therefore are a critical tool for all professional software development.

Closing Comments

This concludes the overview of basic, yet most important, configuration management tools. All tools mentioned above apply just as much to embedded systems projects as they do to all professional software development projects. If you are using configuration control and tracking tools, you appreciate their value; if you are not, your intuition is probably telling you that something is missing. Many good commercial and free tools are available. And they are fun to use, if implemented correctly, because you intuitively know that you are doing the right thing to improve your source code and productivity! Finally note that configuration management is not a destiny, but a continuously evolving process.

I would like to thank Mary Hillstrom for all of her loving support and Bill Littman for his generous advice and for proofreading this paper. Comment? Feel free to send me an email.

Suggested Reading, References, and Resources

AnitPatterns and Patterns in Software Configuration Management
Book by William J. Brown, Hays W. McCormick III, Scott W. Thomas.
Wiley, 1999. Available at http://www.wiley.com/compbooks/catalog/32929-0.htm.

Open Source Development with CVS
Book by Karl Fogel, Coriolis, November 1999.

An outstanding book describing both Open Source Development and the free Concurrent Version System (CVS) used by many (free) large scale and distributed software development teams. CVS is the successor of the well known RCS VC system. Karl convincingly make the case that the two are intimately related. The CVS specific chapters of his book are copyrighted under the GNU General Public License and are available at http://cvsbook.red-bean.com. Available at http://www.coriolis.com/bookstore (search for Fogel).

Code Complete, A Practical Handbook of Software Construction
Book by Steve McConnell, Microsoft Press 1993.

This book is a classic and the books, title says it all, very comprehensive and still in print. Available at http://mspress.microsoft.com/books/28.htm.

"The Version Control Process"
Article by Aspi Havewala, *Dr. Dobb's Journal*, May 1999, page 100.

This article includes a section labeled "Finding a Home for Shared Code" illustrating a nice way to organize shared source code and C/C++ style header files. Available at http://www.ddj.com/articles/1999/9905/9905toc.htm.

"Join The Team"
Article by Alan Radding, *InformationWeek*, October 4, 1999, page 1a.

A recent high level write-up on CM issues and available tools for parallel development and geographically dispersed teams. This is the kind of (rare) article I would like to see many more of in the trade rags. Available at http://www.informationweek.com/755/55adtea.htm.

WinDiff

Software tool by Microsoft Corporation. WinDiff is included with the "Windows NT 4.0 Resource Kit Support Tools" that you can download for free. For more information and the download location search the Microsoft Knowledge Base for article Q206848 at http:// support.microsoft.com/support.

AutoSave

Software tool by Innovative Software. For more information on this inexpensive tool check its home page at http://www.innovativesoftware.com/ and for their distribution partner check http://www.v-com.com/.

comp.software.config-mgmt

News group carried by all traditional news servers. This newsgroup is very active and features up to date frequently asked questions (FAQ). It is intended to be a forum for discussions related to CM both the bureaucratic procedures and the tools used to implement CM strategies. The FAQs is also available at http://www.faqs.org/ if your ISP does not provide a news server or if you do not otherwise have access to a news reader.

microsoft.public.visual.sourcesafe

News group, carried only by the vendor specific news server at "msnews.microsoft.com". This newsgroup is dedicated to Microsoft's Visual Source Safe product. This newsgroup does not appear to have someone maintaining a frequently asked questions (FAQ) list nor does it seem to have active MVPs. Briefly, MVPs are recognized by Microsoft for their willingness to share their knowledge on a peer-to-peer basis. MVPs never seem to tire answering posts. However, most message posts do receive multiple answers within a few hours.

borland.public.teamsource

News group, carried only by the vendor specific news server at "forums.inprise.com". This newsgroup is dedicated to Borland's (or Inprise's) TeamSource product. This newsgroup does not appear to have someone maintaining a frequently asked questions (FAQ) list. This newsgroup is very active and monitored by Borland personal and TeamB members. Briefly, TeamB members are unpaid volunteer professionals selected by Borland, they are not Borland employees, and they tend to be outspoken with their opinions on virtually all aspects of Borland's presence in the market, including but not limited to product development, marketing, and support. TeamB members never seem to tire answering posts.

Math

Introduction

Jack Ganssle

This section contains some useful information for those of us who write code like a=3+b. Everyone, in other words.

Greg Massey's "An Introduction to Machine Calculations" is an important chapter, one with concepts every embedded programmer needs to understand. We treat our compilers as black boxes, assuming they know what we want to do... when usually they don't. What a fun language: 20,000 plus 20,000 is a negative number (for an int)! The math going on behind the scenes is tricky; overflows, underflows and other effects can greatly skew your expected results.

My contributions offer a different look at some floating point and integer operations. Never assume that the only way to solve a problem lies in exploiting the compiler's runtime library. Consider this curve:

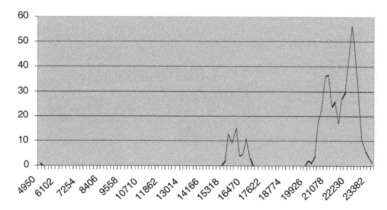

This is a graph of the execution time of a commercial compiler's execution of the cosine routine. The horizontal axis is time. Notice that the time varies all over the map! If we're building a real-time system, it must be both timely, and predictable. Using a function like this means we must always assume worst-case execution time.

There are other algorithms that offer different speed/precision tradeoffs. Most are more predictable.

Choose your algorithms carefully, but don't despair if the vendor's library functions aren't adequate. Plenty of alternatives exist.

An Introduction To Machine Calculations

Gregory W. Massey

Greg Massey is an embedded software engineer with Acterna's Cable Networks Division. Computers have occupied him professionally for over 15 years, during which time he has dealt with test and measurement, communications, automotive, security, military, and factory automation applications. He discovered computers while completing his MS in math at Clemson University. His non-digital interests include his family, his church, and playing classical music (sax and viola). He can be reached at greg.massey@acterna.com.

Introduction

Improvements in languages, compilers, hardware, and even standards continue to simplify the task of performing mathematical calculations in embedded systems. Even with these improvements, though, the calculations do not always perform as expected.

This chapter compares the different types of machine calculations—integer, fixed point, and floating point—in terms of versatility, accuracy and speed. It looks at the sources of errors for each type and ways to manage them. It covers type conversions, overflow effects, implementation tricks, custom integer sizes, and machine dependencies. It shows how to implement fixed-point math using integers and customize it to your needs. The emphasis is on simplicity and practicality. Examples are given in C, although the concepts apply equally well to other programming languages.

Integer Arithmetic

Binary integer arithmetic is as old and familiar as computers themselves. Yet it holds a few surprises that may cause problems for embedded software developers.

Division and Negative Numbers

Consider this simple loop:

```
int i;
for (i = -8; i < 8; ++i)
{
    printf("%3d,", i / 4);
}
```

Integer division is supposed to truncate fractions. We know that 7/4 in integer is not 1 ¾, but 1. The fractional part is discarded. The print statement is dividing all the integers by 4, so we would expect it to print four of each integer value:

```
-2,  -2,  -2,  -2,  -1,  -1,  -1,  -1,   0,   0,   0,   0,   1,   1,   1,   1,
```

But instead it prints one of the first value and 7 zeros:

```
-2,  -1,  -1,  -1,  -1,   0,   0,   0,   0,   0,   0,   0,   1,   1,   1,   1,
```

The difference is caused by the rule that integer division uses for negative results. When it computes –7/4, it does not *truncate downward* to –2 but rather *truncates toward zero*, giving –1.

So why is this important? Suppose the division operation is part of a scaling conversion in a digital thermometer. The input from an analog-to-digital converter must be scaled to the appropriate units for display, perhaps Celsius or Fahrenheit degrees. Divisions between A-D converter counts don't fall exactly on degree boundaries, and the counts contain more precision than needs to be displayed. The scaling operation should give equal weight to all output values, but as the figure below shows, it does not:

| Ideal Output Scaling | -3 | -2 | -1 | 0 | 1 | 2 | 3 |
| Input from A-D Converter |
| Scaled Output for Display | -2 | -1 | | 0 | 1 | 2 | 3 |

The scaling operation converts each count to the nearest whole degree for display. Sometimes three counts get converted to the same output, and sometimes four counts do. But if the scaling algorithm uses integer division that truncates toward 0, there may be twice as many counts that output a 0 as there are for other values.

This bias toward 0 can be a serious problem for any measuring device that outputs both positive and negative values. It gets even worse when we want to divide with rounding instead of truncation. Suppose an input integer represents millivolts and we want to divide by 1000 to get volts. This table shows what we expect and what we get when we truncate toward 0:

Input range	Desired Output	Add 500 for rounding	+1000 Before Truncation	Truncate Toward 0	Truncate Downward
–2500 to –1501	–2	–2000 to –1001	–2.000 to –1.001	–1*	–2
–1500 to –501	–1	–1000 to –1	–1.000 to –0.001	0*	–1
–500 to 499	0	0 to 999	0.000 to 0.999	0	0
500 to 1499	1	1000 to 1999	1.000 to 1.999	1	1
1500 to 2499	2	2000 to 2999	2.000 to 2.999	2	2

The lowest value in the range is not truncated toward 0.

Values of –2.499 volts get reported as –1 volt. Such bias for negative numbers is unacceptable. We can avoid the problem by truncating downward instead of toward 0, but how?

In the print loop above there is an easy solution. A common trick in low-level programming is to shift an integer right by n bits instead of dividing it by 2^n. Shifting an integer value to the right is equivalent to division for positive values and for 0. However, *when a negative value is shifted to the right, it behaves like division that truncates downward.* So rather than divide by 4, we can shift the value right two places.

This trick enables us to divide by a power of two. We need a better solution, though, to divide by other integer values. One way is to use conditional code to handle negative numbers differently. It has disadvantages, though. Conditional code adds another branch that should be tested. It slows down execution, especially if an instruction cache is used. And it adds variability to the timing—a factor that is sometimes critical in embedded software.

Another way is to determine the lowest (negative) number expected, add the negative of it (a positive number) to all inputs, divide, then subtract the result of dividing the offset by the same quotient. For example, suppose the voltmeter described above must handle inputs between –10,000 mV and 10,000 mV. We want to divide by 1000 and display the output in volts. Perform the following steps:

1. Add 10,000 to the input to shift its range up to 0 to 20,000.

2. Add 500 to the result of step 1 so that the result will be rounded rather than truncated.

3. Divide the result of step 2 by 1000. Now the value is between 0 and 20.

4. Subtract 10 from the result of step 3 to get the correctly-rounded voltage.

In step one we use a signed integer for the input. It must be large enough to handle the maximum expected range. In this case it is –10,000 to 10,000. A two-byte integer, which has a range of –32,768 to 32,767, will suffice. After adding the offset, the lowest possible value is 0 and the largest is 20,000. The same two-byte integer can handle this extra range, but for some problems a larger integer type is needed for the result.

In step two we add half the divisor. This step provides the rounding. This step works correctly even if the divisor is an odd number. The result cannot exceed 20,500, so we can still use a two-byte signed integer.

In step three we perform normal integer division. It cannot truncate a negative number toward zero because the offsets in the first two steps guarantee that we are always using a positive value.

In step four we remove the offset. Since the input was divided by 1000, the offset that we subtract must also be 1/1000 of the original offset.

The offset used in steps 1 and 4 does not have to be exactly the same as the largest negative value expected; it can be more, as long as it does not overflow the integer type being used. In this way a single division function can be written to handle many cases.

Truncation toward zero also occurs when a floating-point value is converted to integer. In C and C++ you can safely round a floating point value (*float* or *double*) to the nearest integer using this conversion:

```
int IntVal = (int) floor(FloatVal + 0.5);
```

Add 0.5 so that values of ½ or greater will round up to the next higher integer. The *floor* function returns a value of type *double* that is the greatest integer less than or equal to the input.

Integer Types and Sizes

Here are the typical sizes of the integer data types provided by most C and C++ compilers for microprocessors:

Data Type	Number of Bits	Range of Signed Type Minimum	Range of Signed Type Maximum	Range of Unsigned Type Minimum	Range of Unsigned Type Maximum
char	8	−128	127	0	255
short	16	−32,768	32,767	0	65,535
long	32	−2,147,483,648	2,147,483,647	0	4,294,967,295
custom	n	-2^{n-1}	$2^{n-1}-1$	0	2^n-1

Type *int* is required to be at least 16 bits. It is usually 16 bits for 8-bit and 16-bit processors and 32 bits for 32-bit processors. It is often the best type to use where size is not critical because the compiler uses the size for *int* that it can implement most efficiently.

The last line in the table allows you to compute the exact number of bits needed for special situations. For example, you may need to send large arrays of integer data over a data link. You might be able to improve system throughput significantly by using a custom integer size just big enough for the actual range needed. You can implement it using bit fields in C and C++ or by using bit shifting and masking operations.

Another aspect of integer type is the byte ordering. When microprocessors were developed, some companies chose to make put the low byte of an integer first and others chose to put the high byte first. The common names for these formats are *little-endian* and *big-endian*, respectively.

Integer byte ordering does not become an issue unless data is transferred between processors with different integer byte ordering or data on one processor is alternately interpreted as integer data and as some other type. Copy, equality test, and bitwise operations work with either type unless type promotion (*short* to *long*, etc.) occurs. Other operations require the integer to be reordered into the processor's native type.

Overflow and Underflow

Integer overflow and underflow occur when a computed result exceeds the maximum or the minimum values that can be stored in the data type. On most platforms it does not generate an exception or warning. Instead, it wraps. The most significant bit or bits get discarded and only the bits that will fit into the data type are kept. All of the bits that are kept are correct. Expressed mathematically, the wrapped result W can be calculated from the expected result E using the equation

$$W = E - 2^n \times k$$

where n is the number of bits stored and k is the integer value required to put W within the range of the data type. The next table shows how one-byte integers behave in various overflow and underflow conditions:

Scenario	*unsigned char* Behavior	*signed char* Behavior
Incrementing	... 254, 255, 0, 1, ...	... 126, 127, –128, –127, ...
Decrementing	... 1, 0, 255, 254, ...	... –127, –128, 127, 126, ...
Increasing by 100	0, 100, 200, 44, 144, ...	0, 100, –56, 44, –112, ...
Multiplying by 10	1, 10, 100, 232, 16, 160, ...	1, 10, 100, –24, 16, –96, ...

Programmers sometimes even write software to take advantage of integer wrapping. Consider a system that schedules events for future processing. It has a timer that increments an integer periodically. When an event is scheduled, it gets an integer timestamp at which it is supposed to be processed. Since this is an embedded system, it must be designed to handle timer overflow correctly. The following design takes advantage of integer overflow properties and works for any event delays less than half the clock's maximum count.

The timer value is stored in an unsigned integer *CurrentTime*. It starts at 0, increments periodically, and because it is a good embedded system that runs forever without any problems, the timer eventually overflows and wraps back to 0.

When the system needs to schedule an event to occur some time later, it computes

```
EventTime = CurrentTime + EventDelay
```

using unsigned integers. Sometimes the current time plus the event delay overflow and wrap, giving an event time that is less than the current time.

The system inserts the new event into an ordered list of events waiting to be processed. It needs to put the new event after scheduled events that occur earlier and before those that occur later. The only problem is that an event's time may have wrapped. The insertion algorithm must traverse the list using this simple comparison:

```
unsigned int Difference = NewEventTime - CurrentEventTime;
if ((int) Difference > 0)
{
    . . .        // Advance to the next event in the list.
}
else
{
    . . .        // Insert new event into list before current event.
}
```

The design is surprisingly simple. It needs no special conditional code to handle overflows.

To see how it works, suppose we are using 16-bit unsigned integers for times. The maximum value that can be stored is 65,535. At time 63,000 the system must schedule an event with a 2000-tick delay. Its scheduled processing time is 65,000. The list already contains the following events:

Position	NewEventTime	CurrentEventTime	Expected Difference	Difference	Difference (int)
1	65,000	64,500	500	500	500
2	65,000	600	64,400	64,400	−1136

The system correctly inserts the new event after position 1 but before position 2.

At time 64,000 an event gets scheduled with a 2000-tick delay. Its computed event time is not 66,000, but the wrapped result: 66,000 − 65,536 = 464. The list now contains three items with these times:

Position	NewEventTime	CurrentEventTime	Expected Difference	Difference	(int) Difference
1	464	64,500	−64,036	1500	1500
2	464	65,000	−64,536	1000	1000
3	464	600	−136	65,400	−136

Again, casting the difference to signed int for comparison enables the system to insert the new event at the correct location. This time the new item is inserted after position 2.

Does the cast to *int* just move the overflow problem from the unsigned overflow point to the signed overflow point? Consider the system at time 30,000. It needs to insert a new event with a delay of 3000, giving a new event time of 33,000. The list already contains three items, the first of which has been waiting five ticks for processing.

Position	NewEventTime	CurrentEventTime	Expected Difference	Difference	(int) Difference
1	33,000	29,995	3005	3005	3005
2	33,000	32,900	100	100	100
3	33,000	33,001	−1	65,535	−1

The algorithm will insert the new item after the second position. Again casting the difference to signed integer enables the system to find the correct insertion point.

The system must periodically check to see if the first event in the list needs to be processed. There may be more than one event in the list with the same time, or the system may have missed the exact scheduled processing time, so it uses this loop to process all events older than or equal to the current time:

```
unsigned int Difference = CurrentTime - FirstEventTime;
while ((int) Difference >= 0)
{
    . . .     // Remove and process the first event in
              // the list
    Difference = CurrentTime - FirstEventTime;
}
```

The design does not even need to switch between signed and unsigned types. It can process all times and differences as signed integers. It seems easier to understand, though when times are unsigned.

Floating-Point Math

An Unexpected Result

Floating-point math is a very powerful tool with some capabilities that we would not want to duplicate in integer or fixed-point math, but it also has some problems. If we understand its subtleties, we can work around these problems.

Consider the following loop:

```
float x = 0.0; float dx = 0.1;
int i = 0;
while (x != 1.0)
{
    x += dx;
    i += 1;
}
```

It looks like the loop should iterate 10 times and exit with $x = 1.0$ and $i = 10$. Unfortunately, it loops forever. If you were to examine x after 10 iterations, you would see a value of

1.00000011920. The loop test requires iteration to continue until x has a value of exactly 1.00000000000. This explains the infinite loop.

This looks like a problem of limited precision. Variables of type *float* occupy 4 bytes each. Could we solve the problem by changing variables *x* and *dx* to type *double*, which takes 8 bytes per variable? Unfortunately, no. The program still iterates forever. After 10 iterations, *x* has a value of 0.999999999999999890. The value is more precise; it is good to 15 decimal places instead of 6. But it is still not exact.

A programmer working with type *float* might change the loop test operator from != (not equal) to < (less than) and the problem would be solved. But this is not a general solution; it would not iterate the correct number of times with type *double*. It would also fail in similar situations with other values of *dx*. A better solution to this problem is to change the loop test to x < 0.95. This works for both types *float* and *double*. We expect *x* to be 0.9 on the 9th iteration and 1.0 on the 10th, so this solution works for any errors in *x* of less than 0.05. The errors we saw were much less than this.

In this example the problem is easy to solve, but it raises several questions:

- Where did the errors come from?

- How big can such errors be?

- How can we predict when such errors will occur?

- How can we design solutions to work around them?

Floating Point Formats

In order to answer the first question we need to look at the way floating point numbers are represented on a computer. Most floating point implementations today conform to IEEE 754, the standard for binary floating point math. This standard defines two sizes of floating point formats and minimum requirements for two additional sizes.

A floating point number has three parts: a sign bit, a mantissa, and an exponent. Here are the IEEE 754 standard formats:

	sign	exponent	mantissa
float	1bit	8 bits	23 bits
double	1bit	11 bits	52 bits

The sign bit is 0 for positive numbers and 1 for negative numbers. (Zero is a special case; there are different representations for +0 and –0, but they must be treated as equal.)

The mantissa and exponent work about like they do in scientific notation. The mantissa has a binary value and the exponent tells how many places the mantissa is shifted to the left or right. They are not exactly like scientific notation for two reasons. First, they are done in binary instead of decimal, so each digit occupies a "binary place" rather than a decimal place.

There is a "binary point" instead of a decimal point that separates the whole number part from the fractional part. Second, the numbers stored in the exponent field and the mantissa field don't exactly represent the actual exponent and mantissa.

Actual mantissa values come in two forms: normalized and denormalized. The normalized form is used for all but very small numbers. The exponent is chosen so that the whole number part is one. (Compare this to scientific notation in which there is one nonzero digit before the decimal point.) Since the whole number part always has a value of one, there is no need to store it. Only the fractional part gets stored in the mantissa field.

Actual exponent values may be positive or negative. Rather than use two's complement in the exponent field like signed integers use, the field is always interpreted as an unsigned integer. The standard specifies a bias that must be subtracted from this unsigned value in order to get the signed exponent value. (The bias method allowed the designers to support a greater range of positive exponents. The range of negative exponents is reduced, but they make up for the loss by changing the mantissa format.)

So far we have a system for representing a wide range of positive and negative numbers with both large and small magnitudes. In order to be complete, though, the system needs a zero value. It also needs ways to indicate overflow and a variety of error conditions. An exponent of all zero bits or all one bits means to interpret the mantissa as one of these special conditions. The next table shows the special values and formats defined by the standard:

Sign	Exponent	Mantissa	Meaning
0, 1	00...0	00...0	Zero (+0 or –0)
0, 1	00...0	00...1 through 11...1	Denormalized numbers for representing values with very small magnitudes. The implied whole number portion of the mantissa is 0.
0, 1	00...1 through 11...0	00...0 through 11...1	Normalized numbers for representing the vast majority of floating point values. The implied whole number portion of the mantissa is 1.
0, 1	11...1	00...0	± infinity, depending on sign bit
1	11...1	10...0	Indeterminate ($+\infty + -\infty$ or $\pm\infty \times 0$)
0	11...1	any	NaN (not a number). May represent uninitialized data, etc.
1	11...1	all but 10...0	

Here are more details about the IEEE 754 formats:

Attribute	float	double
Size	4 bytes	8 bytes
Normalized mantissa size (includes implied "1.")	24 bits	53 bits
Normalized mantissa range	1 to $(2 - 2^{-23})$	1 to $(2 - 2^{-52})$
Equivalent decimal precision	over 7 digits	over 15 digits
Exponent size	8 bits	11 bits
Exponent bias	127	1023
Exponent range	-126 to 127	-1022 to 1023
Number range with full precision	$0, \pm 2^{-126}$ to $\approx 2^{128}$	$0, \pm 2^{-1022}$ to $\approx 2^{1024}$
Equivalent decimal range	$0, \pm \approx 1.2 \times 10^{-38}$ to $\approx 3.4 \times 10^{38}$	$0, \pm \approx 2.2 \times 10^{-308}$ to $\approx 1.8 \times 10^{308}$

Now let's see how the value 11 is represented in type *float*. The value is positive, so the sign bit is 0. The whole number part must be 1 and the exponent chosen accordingly. Since $11 = 1.375 \times 2^3$, we need a mantissa value of 1.375 and an exponent value of 3.

To get the mantissa, start with the desired value of 1.375. Subtract the implied 1 to get 0.375. The result is less than 2^{-1}, so put a 0 in the first bit. It is not less than 2^{-2}, so put a 1 in the next bit and subtract 2^{-2} from the desired value, giving a new desired value of 0.125. It is not less than 2^{-3}, so put a 1 in the third bit and subtract 2^{-3}. The desired value is now 0 so we can fill in all remaining bits with 0.

To get the exponent, add the bias of 127 to the desired exponent value of 3. Put the result, 130, into binary to get 10000010.

Here is our *float* representation of 11:

	sign	exponent	mantissa
11 float	0	10000010	01100000000000000000000

Roundoff Errors

Now let's get the representation for 0.1. Normalize to get $0.1 = 1.6 \times 2^{-4}$. Add the bias to the exponent to get 123. Subtract the implied 1 from the mantissa and get 0.6. Subtract 2^{-1} and get 0.1. Subtract 2^{-4} and get 0.0375, then subtract 2^{-5} and get 0.00625, and so on. After subtracting 2^{-23} we still have a value big enough that we could subtract 2^{-24}. We have a problem, though. We have run out of room in our mantissa. The IEEE standard says that we must round the mantissa to the nearest value; in this case that gives us

	sign	exponent	mantissa
0.10000000149 float	0	01111011	10011001100110011001101

which helps us understand why our looping problem occurred. Quite simply, we cannot represent 0.1 exactly in binary floating point. The necessary mantissa is a repeating number in binary; any size floating point representation will truncate it somewhere and give us a value that is slightly different from 0.1.

As just demonstrated, there are numbers that cannot be represented exactly in floating point. The binary representations repeat, even though the decimal representations may not. Generalizing what we just saw, *if any number is converted to a fraction in lowest terms, and the denominator of that fraction is not a power of two, that number will not have an exact floating point representation.* This is one common source of floating point error.

To see a similar type of floating point error, divide 2 by 5. Both 2 and 5 have exact floating point representations, but the result, 0.4, does not. The result given in type float is 0.40000000596. The difference is caused by roundoff error: *A calculated result with no exact floating point representation is converted to the nearest floating point number.* The IEEE 754 standard requires that the computed result of addition, subtraction, multiplication, division, and square root be the nearest floating point number to the theoretical result, so the resulting error will be no more than ½ bit. (Before the standard was introduced, many floating point implementations introduced errors of one bit or more. In order to comply with this requirement, the floating point processor or library must actually perform the calculation with a few extra bits.)

The division described above has an error of 5.96×10^{-9}. How bad is this? The worst possible error for a result of that magnitude (one that requires the same exponent in its *float* representation) is $\pm\frac{1}{2}$ bit in the 23rd place. This is the same as ± 1 bit in the 24th place. The exponent used for numbers between 0.25 and 0.5 is -2 (before the bias is added), so the worst-case error would be $\pm 2^{-26}$ or about $\pm 1.5 \times 10^{-8}$.

Now an error of $\pm 1.5 \times 10^{-8}$ would be bad if the computed result were 10^{-7}. It would be insignficant, though, for a large result like 10^{23}. In other words, expressing error as a number does not tell us enough.

Define the *relative error* to be the absolute error divided by the expected result. Now the relative error in our $2 \div 5$ problem is $(1.5 \times 10^{-8}) / 0.4$, or 3.75×10^{-8}. The representation error for 0.1 has exactly the same amount of relative error. Relative error frees us from having to know the magnitude of the expected result.

The worst-case relative roundoff error for a single type *float* arithmetic calculation is about 6×10^{-8}; for type double it is about 1.1×10^{-16}. These are not bad at all. Not many instruments can measure physical quantities with 7-digit precision, let alone nearly 16-digit precision!

If floating-point errors got no worse than these, life would be very good indeed in the world of machine calculations. But the errors can get much worse. First we will consider cumulative effects of a series of multiplications and divisions. Then we'll look at a much more serious problem that can arise with addition and subtraction.

Effects of Errors in Multiplication and Division

A bit more notation will make the next section easier. Mathematicians and computer scientists define the machine epsilon ϵ as the maximum amount of relative error that would be produced by inverting the least-significant bit of the mantissa. Thus, a single multiplication or division may produce a roundoff error of $\pm\epsilon/2$. A sequence of n multiplications and divisions, each using the result of a previous operation, could produce a cumulative error of $\pm n\epsilon/2$. The cumulative error is usually much less than this because the individual errors do not all have the same sign or the maximum possible magnitude.

Using the cumulative error result above, we can get the number of possible bits in error from

$$\text{Number of suspect bits: } b_S = \text{ceil}(\log_2 n)$$

where n is the number of sequential $\times$ or $\div$ operations, and the function ceil(x) raises non-integer x values to the next integer. Now define

$$\text{Number of good bits: } b_G = b_M - b_S$$

where b_M is the number of bits in the mantissa, including the implied leading bit: 24 for *float*, 53 for *double*.

The number of good base 10 digits is related to the number of good bits in the mantissa by

$$\text{Number of good decimal digits: } d_G = \text{floor}(b_G \times \log_{10} 2)$$

where floor(x) is the integer portion of x. With a little algebra we get the result:

$$\text{Maximum number of sequential } \times \text{ or } \div \text{ operations: } n_{\text{max}} = 2^{(b_M - \text{ceil}(d_G/\log_{10} 2))}$$

Here is a C function that performs the calculation:

```
int MaxMultDiv(int GoodDigits, int MantissaBits)

{
    double Power = MantissaBits;
    Power -= ceil(GoodDigits / 0.30103);
    return (int) (pow(2.0, Power) + 0.5);
}
```

The following table shows some worst-case results for the two common mantissa sizes:

Number of Digits Needed	Max Number of × and ÷ Operations Allowed	
	float	*double*
4	1024	5.5E+11
5	128	6.87E+10
6	16	8.59E+09
12		8192
13		512
14		64
15		8

These are worst-case results for sequential multiplication and division. Your results will probably be much better.

Effects of Errors in Addition and Subtraction

Consider the following code:

```
float pi  = 3.14159265359;
float k   = 1000.0;
float sum = pi + k;

printf("pi  = %15.7f\n", pi);
printf("sum = %15.7f\n", sum);
```

It produces the following output:

```
pi  =      3.1415927
sum = 1003.1416016
```

The second value is obviously inaccurate. Both values, though, are correct to 8 significant digits. The relative error is small. As in multiplication and division, the result is guaranteed to be no worse than $\pm\varepsilon/2$. That still does not seem very satisfying. Perhaps it is because we can mentally add the two values so easily with no errors at all. This simple example shows that *whenever floating-point values with differing magnitudes are added or subtracted, the one with the smaller magnitude loses precision.*

Continuing with this example, let's see what happens when we try to recover the value of *pi*. Computing

```
float pi2 = sum - k;
printf("pi2 = %15.7f\n", pi2);
```

gives us the result:

```
pi2 =      3.1416016
```

This is no surprise, given the value of the sum, but it still points out a problem. Now our computed value is correct to only 5 significant digits because of inaccuracy in an intermediate value.

Let's examine the add and subtract operations more closely. Here are the two values being added, with mantissas shown in binary form:

$3.1415927 = 1.100\ 1001\ 0000\ 1111\ 1101\ 1011 \times 2^1$
$1000 \qquad = 1.111\ 1010\ 0000\ 0000\ 0000\ 0000 \times 2^9$

The exponents must be equal in order to add the two values. If we shift the mantissa right one bit, we can increase the exponent by 1. Shifting right 8 places gives:

$3.1415927 = 0.000\ 0000\ 1100\ 1001\ 0000\ 1111\ 1101\ 1011 \times 2^9$

Now add the two values to get:

1003.1415927 = 1.111 1010 1100 1001 0000 1111 *1101 1011* × 2⁹

The mantissa has too many binary digits. The last 8 digits, shown in italics, must be discarded. According to IEEE 754 the result must be rounded, not truncated, giving:

1003.1416016 = 1.111 1010 1100 1001 0001 0000 × 2⁹

Subtracting 1000 leaves us with:

3.1416016 = 0.000 0000 1100 1001 0001 0000 × 2⁹

We must shift the mantissa left until the first bit is a 1 (normalization). Each shift decreases the exponent by 1. We don't have data to shift into the rightmost places, so we fill them with 0's. The computed result is given below, along with the desired result:

3.1416016 = 1.100 1001 0001 0000 0000 0000 × 2¹
3.1415927 = 1.100 1001 0000 1111 1101 1011 × 2¹

Subtraction of nearly equal values produces *cancellation*. If the two numbers being subtracted are exactly what they are supposed to represent, the result is *benign cancellation*. If, as in this case, at least one of the values is only an approximation, the result is *catastrophic cancellation*.

This principle applies to both positive and negative floating-point numbers. When two numbers are nearly equal in magnitude but have opposite signs, adding them will produce cancellation. If at least one of the values being added is an approximation with limited precision, catastrophic cancellation will result.

In summary, addition and subtraction can produce errors. The seriousness of the errors depends on the signs and magnitudes of the numbers being added or subtracted, and whether they are exact values or approximations. The following tables show all the cases that need to be considered for addition and subtraction:

Addition	Magnitudes	
	Similar	**Different**
Signs — **Same**	Safe; error is no more than ±ε/2.	May lose precision of value with smaller magnitude.
Signs — **Different**	Cancellation can occur.	

Subtraction	Magnitudes	
	Similar	**Different**
Signs — **Same**	Cancellation can occur.	May lose precision of value with smaller magnitude.
Signs — **Different**	Safe; error is no more than ±ε/2.	

Managing Floating Point Errors

Comparing Floating Point Values

If you need to compare a computed result x to a constant c, write the test so that the actual computed value can be within a small neighborhood of the expected value. Establish a tolerance *eps* based on the expected precision of x. For type *float* it might be $c \times 0.00001$. (If c is zero, of course, you cannot multiply to get *eps*.) If you need to compare two computed results x and y, the tolerance may need to be relative to the size of one or the other. Use a relative error coefficient *rel* that is consistent with the precision of the floating-point type you are working with.

Instead of	Use	Comments
if (x == c)	if (fabs(x - c) < eps)	
if (x <= c)	if (x < c + eps)	True if result is close
if (x < c)	if (x < c - eps)	False if result is close
if (x == y)	if (fabs(x - y)< x * rel)	May use either x or y to compute acceptable error

Adding a Sequence of Numbers

Many calculations require a sequence of numbers to be added. Often the problem is stated so that the largest magnitude terms are summed first and the smallest ones are summed last. If you know something about the sequence, you can often program it so that the terms are summed from smallest magnitude to largest.

A better way, especially if the data is not inherently sorted, is to use the Kahan Summation Formula:

```
double KahanSum(double Term[], int NumTerms)
{
    double Sum = Term[0];
    double Adjustment = 0;
    double AdjustedTerm;
    double NewSum;

    for (int i = 1; i < NumTerms; ++i)
    {
        AdjustedTerm = Term[i] - Adjustment;
        NewSum = Sum + AdjustedTerm;
        Adjustment = (NewSum - Sum) - AdjustedTerm;
        Sum = NewSum;
    }
    return Sum;
}
```

Computing Polynomials

Rather than compute a polynomial function such as

$$f(x) = 5x^4 + 3x^3 - 7x^2 + 2x - 12$$

using exponents, compute the equivalent function

$$f(x) = ((5x + 3)x - 7)x + 2)x - 12$$

which gets powers of x through repeated multiplication. It tends to be both faster and more accurate. This little trick is known as *Horner's Rule*.

Alternative Number Formats

Binary floating point is not the best choice for some applications. If it is important to represent numbers such as 0.1 exactly, you can use a decimal floating point system. Each decimal digit in the mantissa is stored in four bits. Such a system is called *binary-coded decimal* (BCD), and if each byte contains two decimal digits, it is called *packed* BCD. Even though packed BCD uses all the bits, it does not use the space as efficiently as binary floating point. For example, type *float* uses 23 stored bits in the mantissa and provides a precision that exceeds 7 decimal digits. A packed BCD mantissa with equivalent precision would need at least 28 stored bits.

Another type of alternative system, *rational arithmetic*, uses an integer for a numerator and another for a denominator. This system can represent any rational number exactly up to the limits of its numerator and denominator ranges. It is well-suited to problems with fractions but not good at trigonometric or logarithmic functions. Unfortunately, implementations of even the basic operations are complex and slow because they have to compute greatest common factors. Because of this, rational arithmetic is seldom used.

Using Equivalent Calculations to Avoid Catastrophic Cancellation

How bad is catastrophic cancellation? In the worst situations it can reduce the result to no significant digits at all! Suppose we need to calculate the value

$$y = 1 - \cos(x)$$

for small values of x. The cosine of 0 is 1, and the cosine of a small value will be very close to 1. If we compute y directly, cancellation will occur when you subtract 1. Is it benign or catastrophic? The mathematical cosine function generally produces irrational results—they cannot be represented exactly by either binary or decimal numbers. Therefore the computed cosine value is an approximation. The result of the subtraction will be catastrophic. For very small values of x, the y values will not even be accurate to one digit. This equation is said to be *numerically unstable* for small values of x.

Can we convert the equation to something equivalent that does not require this catastrophic subtraction? We can use the identity

$$\cos^2 a + \sin^2 a = 1$$

to convert the formula to the equivalent formula:

$$y = \frac{\sin^2(x)}{1 + \cos(x)}$$

Now the catastrophic subtraction goes away for small values of x. The cosine function has a value of about 1, so nearly equal numbers are being added and not subtracted.

This equation, though, has other problems. We have increased the complexity. The trigonometric functions take longer to compute than arithmetic operations, and now we have two of them. Furthermore, x values close to $\pm\,\pi, \pm\,3\pi, \pm\,5\pi, \ldots$ will make the computed cosine value nearly equal to -1. Catastrophic cancellation occurs because nearly opposite values are added. Some values might even cause division by 0.

If we can put up with the extra coputing time, we can use one form of the equation for some x values and the other form for others. We can do better than that, though.

A little more time spent in a math handbook will lead us to this identity:

$$\sin \tfrac{a}{2} = \pm\sqrt{\tfrac{1 - \cos a}{2}}$$

Now we can convert the equation to

$$y = 2\sin^2(x/2)$$

which has no addition or subtraction, no division by small numbers (therefore no chance of dividing by 0), and only one trigonometric function. Now we can get the accuracy we need with practically no performance penalty.

Here are the computed results using both forms of the equation:

Computing y = 1 – cos(x) with original and modified equations

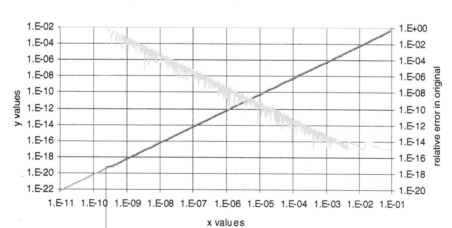

The scales are logarithmic, so the graph does not have the familiar cosine wave shape. Values were computed using IEEE 754 double precision. The modified equation works for all x values in the graph. (It was tested for x values down to 10^{-100} with no problems.) The original equation starts to be inaccurate when x goes below 10^{-3}. At that point the relative error, plotted against the right axis, starts to go up. As x gets smaller, the relative error increases until it finally reaches 1 for x values below about 2×10^{-10}. There the original equation fails completely and returns 0. The modified equation produces results with 14-digit accuracy for all x values. At $x \approx 7\times10^{-7}$ the original equation's relative error of 10^{-7} makes the resulting accuracy only about 7 digits.

This example is not contrived to exaggerate the benefits of avoiding catastrophic cancellation. It is typical of the gains in precision that can be obtained by using mathematical identities to transform numerically unstable formulas into numerically stable ones.

Not all numerically-stable equivalent formulas are simple or obvious. This one from Hewlett Packard computes $y = \ln(1 + x)$ accurately for small values of x:

$$y = \begin{cases} x & \text{for} \quad 1 \oplus x = 1 \\ \dfrac{x\ln(1+x)}{(1+x)-1} & \text{for} \quad 1 \oplus x \neq 1 \end{cases}$$

The symbol $\oplus$ denotes machine addition with roundoff errors.

Fixed-Point Arithmetic

When Should It Be Used?

Some types of problems require fractional values but do not really need floating point. They can be done in fixed point, which is really just integer math accompanied by some rules and conventions.

Here are some criteria for determining whether fixed point would be better than floating point:

- The values operate over a limited range.

- Values near 0 do not need to have greater resolution than high-magnitude values.

- The calculations are linear (multiplying or dividing by constants, addition and subtraction, and interpolation).

- There are tight speed or memory space requirements.

- The platform provides no hardware support for floating point.

Not all have to be true in order for fixed point to be a better alternative than floating point. They are listed in roughly descending order of significance—fixed point may be the better choice for an application even if the processor has a floating-point coprocessor.

Fixed-point Number Representation and Operations

Fixed point math is implemented using integers and a scale factor. Instead of having the least-significant bit represent one unit, it represents a unit divided by a scale factor. For example, a thermometer with 0.1 degree resolution would use a scale factor of 10 to make the least significant bit represent tenths of a degree.

To represent a real number x in fixed point with scale factor s, use the closest integer value:

$$i_x = \text{floor}(s \cdot x + 0.5)$$

The scale factor s is usually greater than 1, but it does not have to be.

The following table provides fixed-point implementations for common operations. A sample problem is provided for each operation. The values used for the sample problems are shown in the second table.

Operation	Definition	Implementation (See note below)	Sample Values		
			Desired	Integer	Represented
Addition	$x + y$	$n_x + n_y$	112.5	1125	112.5
Subtraction	$x - y$	$n_x - n_y$	82.3	823	82.3
Multiplication	$x \cdot y$	$\frac{(n_x \cdot n_y + s/2)}{s}$	1470.74	14707	1470.7
Division	x/y	$\frac{(s \cdot n_x + n_y/2)}{n_y}$	6.45033	65	6.5
Interpolation	$(a \cdot x + b \cdot v)/(a + b)$	$\frac{(a \cdot n_x + b \cdot n_y + (a+b)/2)}{a + b}$	31.56	316	31.6

Note: The final division in the last three operations must be done with truncation downward, not truncation toward zero, in order to handle negative values correctly.

In the sample values section, the desired column is the ideal result that we should get if precision were not limited. The integer column shows the integer value obtained from the formula. The Represented column shows the fixed-point value represented by the integer value.

Values used in sample problems:

Variable	Value	Description
s	10	Fixed point scale factor
x	97.4	Input values used to demonstrate how each operation is performed
y	15.1	
a	1	Coefficients used in interpolation formula
b	4	

The fixed-point implementations of addition and subtraction are exact. Those for multiplication and division must round the result to the nearest integer in the implementation. The formulas given round all results correctly. They do not include tests for overflow or underflow. If possible, the integer size should be chosen so that such tests are not needed.

Managing Errors in Fixed-point Math

When designing with fixed-point math, it is important to choose scale factors carefully. It is especially dangerous to mix fixed-point systems with different scale factors, as the following example shows:

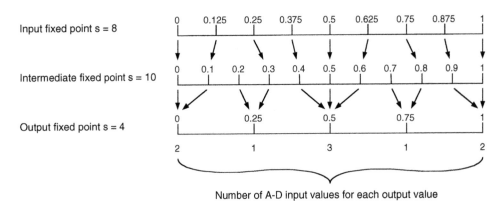

Number of A-D input values for each output value

The input values have one scale factor, the intermediate values have another, and the output values have yet another. By tracing each possible input value, we see that some output values are used for only one input value each, but one output value represents three different input values. A linear change in input will not produce a linear change in output.

In order to avoid this problem, when one scale factor is given by system requirements, any subsequent scale factors introduced by the design should be multiples of it. When two scale factors are required (such as input and output scale factors), any intermediate scale factors should be multiples of both. That means the least acceptable intermediate scale factor is the least common multiple of the two required scale factors.

Addition and subtraction can be done exactly, but multiplication and division cannot. One way to minimize cumulative errors is to use an intermediate fixed-point scale factor that provides a few extra decimal places. A longer integer type may be needed for the values with the larger scale factor.

Conclusion

Each of the common forms of machine arithmetic—integer, floating point, and fixed point—has advantages and disadvantages. Software developers need a practical understanding of each of them in order to choose the best form for an application and avoid the problems associated with the form being used.

This chapter provides only an overview into the topic of numerical analysis. If a more thorough analysis is needed for specialized calculations such as matrix operations, a numerical analysis textbook will provide much more detail.

Bibliography

David Goldberg, *What Every Computer Scientist Should Know About Floating Point Arithmetic*, Computing Surveys, March 1991.

IEEE Computer Society (1985), *IEEE Standard for Binary Floating-Point Arithmetic*, IEEE Std 754-1985.

Peter A. Stark, *Introduction to Numerical Methods,* Macmillan Publishing Co., 1970.

Josef Stoer, *Introduction to Numerical Analysis*, Springer-Verlag, 1980.

CHAPTER **12**

Floating Point Approximations

Jack Ganssle

Most embedded processors don't know how to compute trig and other complex functions. When programming in C, we're content to call a library routine that does all of the work for us. Unhappily this optimistic approach often fails in real-time systems where size, speed and accuracy are all-important issues.

The compiler's runtime package is a one-size-fits-all proposition. It gives a reasonable trade-off of speed and precision. But every embedded system is different, with different requirements. In some cases it makes sense to write our own approximation routines. Why?

> *Speed*—Many compilers have very slow runtime packages. *A clever approximation may eliminate the need to use a faster CPU.*

> *Predictability*—Compiler functions vary greatly in execution time depending on the input argument. Real-time systems must be *predictable* in the time domain. The alternative is always to assume worst case execution time, which again may mean your CPU is too slow, too loaded, for the application.

> *Accuracy*—Read the compiler's manuals carefully! Some explicitly do not support the ASNI C standard, which requires all trig to be double precision. (8051 compilers are notorious for this). Alternatively, why pay the cost (in time) for double precision math when you only need 5 digits of accuracy?

> *Size*—When memory is scarce, using one of these approximations may save much code space. If you only need a simple cosine, why include the entire floating point trig library?

This collection is not an encyclopedia of all possible approximations; rather, it's the most practical ones distilled from the bible of the subject, *Computer Approximations* by John Hart (ISBN 0-88275-642-7). Unfortunately this work is now out of print. It's also very difficult to use without a rigorous mathematical background.

All of the approximations here are polynomials, or ratios of polynomials. All use very cryptic coefficients derived from Chebyshev series and Bessel functions. Rest assured that these numbers give minimum errors for the indicated ranges. Each approximation (with a few exceptions) has an error chart so you can see exactly where the errors occur. In some cases, if you're using a limited range of input data, your accuracies can far exceed the indicated values. For instance, cos_73 is accurate to 7.3 decimal digits over the 0 to 360 degree range.

But as the graph shows, in the range 0 to 30 degrees you can get almost an order of magnitude improvement.

Do be wary of the precision of your compiler's floating point package. Some treat `doubles` as `floats`. Others, especially for tiny CPUs like the PIC, cheat and offer less than full 32-bit floating point precision.

All of the code for the following approximations was compiled with Microsoft's Visual C++ 6.0. The source is available at www.ganssle.com/approx/sincos.cpp. It includes test code that writes a text file of results and errors; imported to a spreadsheet we can see just how accurate they are.

General Trig Notes

We generally work in radians rather than degrees. The 360 degrees in a circle are equivalent to 2π radians; thus, one radian is $360/(2\pi)$, or about 57.3 degrees. This may seem a bit odd till you think of the circle's circumference, which is $2\pi r$; if r (the circle's radius) is one, the circumference is indeed 2π.

The conversions between radians and degrees are:

```
Angle in radians= angle in degrees * 2 π /360

Angle in degrees= angle in radians * 360/(2 π)
```

Degrees	Radians	Sine	Cosine	Tangent
0	0	0	1	0
45	$\pi/4$	$\sqrt{2}/2$	$\sqrt{2}/2$	1
90	$\pi/2$	1	0	infinity
135	$3\pi/4$	$\sqrt{2}/2$	$-\sqrt{2}/2$	−1
180	π	0	−1	Infinity
225	$5\pi/4$	$-\sqrt{2}/2$	$-\sqrt{2}/2$	1
270	$3\pi/2$	−1	0	Infinity
315	$7\pi/4$	$-\sqrt{2}/2$	$\sqrt{2}/2$	−1
360	2π	0	1	0

Cosine and Sine

The following examples all approximate the cosine function; sine is derived from cosine via the relationship:

$$\sin(x) = \cos(\pi/2 - x)$$

In other words, the sine and cosine are the same function, merely shifted 90° in phase. The sine code is (assuming we're calling cos_32, the lowest accuracy cosine approximation):

```
//      The sine is just cosine shifted a half-pi, so
// we'll adjust the argument and call the cosine approximation.
//
float sin_32(float x){
                return cos_32(halfpi-x);
}
```

All of the cosine approximations in this chapter compute the cosine accurately over the range of 0 to π/2 (0 to 90°). That surely denies us of most of the circle! Approximations in general work best over rather limited ranges; it's up to us to reduce the input range to something the approximation can handle accurately.

Therefore, before calling any of the following cosine approximations we assume the range has been reduced to 0 to π/2 using the following code:

```
// Math constants
double const pi=3.1415926535897932384626433;// pi
double const twopi=2.0*pi;      // pi times 2
double const halfpi=pi/2.0;     // pi divided by 2
//
//  This is the main cosine approximation "driver"
// It reduces the input argument's range to [0, pi/2],
// and then calls the approximator.
//
float cos_32(float x){
        int quad;                       // what quadrant are we in?

        x=fmod(x, twopi);               // Get rid of values > 2* pi
        if(x<0)x=-x;                    // cos(-x) = cos(x)
        quad=int(x/halfpi);             // Get quadrant # (0 to 3)
        switch (quad){
        case 0: return  cos_32s(x);
        case 1: return -cos_32s(pi-x);
        case 2: return -cos_32s(x-pi);
        case 3: return  cos_32s(twopi-x);
        }
}
```

This code is configured to call `cos_32s`, which is the approximation (detailed shortly) for computing the cosine to 3.2 digits accuracy. Use this same code, though, for all cosine approximations; change cos_32s to cos_52s, cos_73s or cos_121s, depending on which level of accuracy you need. See the complete listing for a comprehensive example.

Be clever about declaring variables and constants. Clearly, working with the cos_32 approximation nothing must be declared "double". Use float for more efficient code. Reading the complete listing you'll notice that for cos_32 and cos_52 we used floats everywhere; the more accurate approximations declare things as doubles.

One trick that will speed up the approximations is to compute x^2 by incrementing the characteristic of the floating point representation of x. You'll have to know exactly how the numbers are stored, but can save hundreds of microseconds over performing the much clearer "x*x" operation.

How does the range reduction work? Note that the code divides the input argument into one of four "quadrants"—the very same quadrants of the circle shown below:

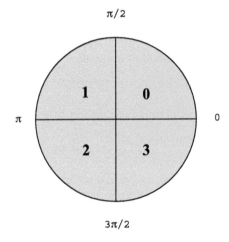

Figure 12-1: Quadrants 0 to 3 of the circle.

- For the first quadrant (0 to π/2) there's nothing to do since the cosine approximations are valid over this range.

- In quadrant 1 the cosine is symertrical with quadrant 0, if we reduce it's range by subtracting the argument from π. The cosine, though, is negative for quadrants 1 and 2 so we compute −cos(π−x).

- Quadrant 2 is similar to 1.

- Finally, in 3 the cosine goes positive again; if we subtract the argument from 2 π it translates back to something between 0 and π/2.

The approximations do convert the basic polynomial to a simpler, much less computationally expensive form, as described in the comments. All floating point operations take appreciable amounts of time, so it's important to optimize the design.

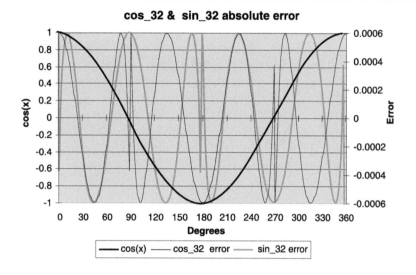

```
//                     cos_32s computes cosine (x)
//
//   Accurate to about 3.2 decimal digits over the range [0, pi/2].
//   The input argument is in radians.
//
//   Algorithm:
//                cos(x)= c1 + c2*x**2 + c3*x**4
//     which is the same as:
//                cos(x)= c1 + x**2(c2 + c3*x**2)
//
float cos_32s(float x)
{
const float c1= 0.99940307;
const float c2=-0.49558072;
const float c3= 0.03679168;

float x2;                              // The input argument squared

x2=x * x;
return (c1 + x2*(c2 + c3 * x2));
}
```

cos_32 computes a cosine to about 3.2 decimal digits of accuracy. Use the range reduction code (listed earlier) if the range exceeds 0 to $\pi/2$. The plotted errors are absolute (not percent error).

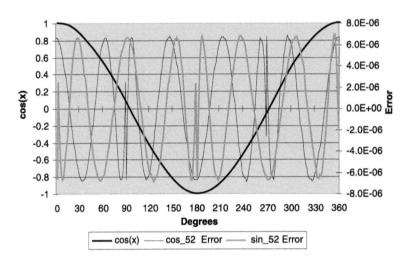

cos_52 & sin_52 absolute error

```
//                     cos_52s computes cosine (x)
//
//   Accurate to about 5.2 decimal digits over the range [0, pi/2].
//   The input argument is in radians.
//
//   Algorithm:
//                     cos(x)= c1 + c2*x**2 + c3*x**4 + c4*x**6
//     which is the same as:
//                     cos(x)= c1 + x**2(c2 + c3*x**2 + c4*x**4)
//                     cos(x)= c1 + x**2(c2 + x**2(c3 + c4*x**2))
//
float cos_52s(float x)
{
const float c1= 0.9999932946;
const float c2=-0.4999124376;
const float c3= 0.0414877472;
const float c4=-0.0012712095;
float x2;                                // The input argument squared

x2=x * x;
return (c1 + x2*(c2 + x2*(c3 + c4*x2)));
}
```

cos_52 computes a cosine to about 5.2 decimal digits of accuracy. Use the range reduction code (listed earlier) if the range exceeds 0 to $\pi/2$. The plotted errors are absolute (not percent error).

cos_73 & sin_73 absolute error

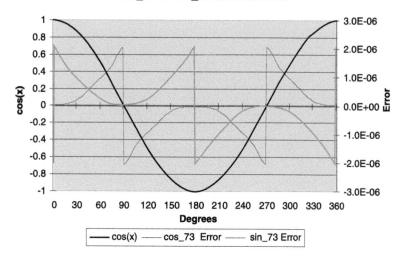

```
//                    cos_73s computes cosine (x)
//
//   Accurate to about 7.3 decimal digits over the range [0, pi/2].
//   The input argument is in radians.
//
//   Algorithm:
//                cos(x) = c1 + c2*x**2 + c3*x**4 + c4*x**6 + c5*x**8
//      which is the same as:
//                cos(x) = c1 + x**2(c2 + c3*x**2 + c4*x**4 + c5*x**6)
//                cos(x) = c1 + x**2(c2 + x**2(c3 + c4*x**2 + c5*x**4))
//                cos(x) = c1 + x**2(c2 + x**2(c3 + x**2(c4 + c5*x**2)))
//
double cos_73s(double x)
{
const double c1= 0.999999953464;
const double c2=-0.4999999053455;
const double c3= 0.0416635846769;
const double c4=-0.0013853704264;
const double c5= 0.00002315393167;
double x2;                               // The input argument squared

x2=x * x;
return (c1 + x2*(c2 + x2*(c3 + x2*(c4 + c5*x2))));
}
```

cos_73 computes a cosine to about 7.3 decimal digits of accuracy. Use the range reduction code (listed earlier) if the range exceeds 0 to $\pi/2$. Also plan on using double precision math for the range reduction code to avoid losing accuracy. The plotted errors are absolute (not percent error).

cos_121 & sin_121 absolute error

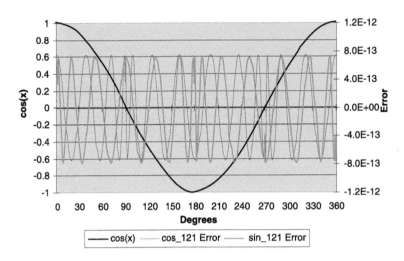

```
//                    cos_121s computes cosine (x)
//
//   Accurate to about 12.1 decimal digits over the range [0, pi/2].
//   The input argument is in radians.
//
//   Algorithm:
//        cos(x) = c1+c2*x**2+c3*x**4+c4*x**6+c5*x**8+c6*x**10+c7*x**12
//   which is the same as:
//        cos(x) = c1+x**2(c2+c3*x**2+c4*x**4+c5*x**6+c6*x**8+c7*x**10)
//        cos(x) = c1+x**2(c2+x**2(c3+c4*x**2+c5*x**4+c6*x**6+c7*x**8 ))
//        cos(x) = c1+x**2(c2+x**2(c3+x**2(c4+c5*x**2+c6*x**4+c7*x**6 )))
//        cos(x) = c1+x**2(c2+x**2(c3+x**2(c4+x**2(c5+c6*x**2+c7*x**4 ))))
//        cos(x) = c1+x**2(c2+x**2(c3+x**2(c4+x**2(c5+x**2(c6+c7*x**2 )))))
//
double cos_121s(double x)
{
const double c1= 0.99999999999925182;
const double c2=-0.49999999997024012;
const double c3= 0.041666666473384543;
const double c4=-0.0013888888418000423;
const double c5= 0.0000248010406484558;
const double c6=-0.0000002752469638432;
const double c7= 0.0000000019907856854;
double x2;                                // The input argument squared

x2=x * x;
return (c1 + x2*(c2 + x2*(c3 + x2*(c4 + x2*(c5 + x2*(c6 + c7*x2)))))); 
}
```

cos_121 computes a cosine to about 12.1 decimal digits of accuracy. Use the range reduction code (listed earlier) if the range exceeds 0 to $\pi/2$. Also plan on using double precision math for the range reduction code to avoid losing accuracy. The plotted errors are absolute (not percent error).

Higher Precision Cosines

Given a large enough polynomial there's no limit to the possible accuracy. A few more algorithms are listed here. These are all valid for the range of 0 to $\pi/2$, and all can use the previous range reduction algorithm to change any angle into one within this range. All take an input argument in radians.

No graphs are included because these exceed the accuracy of the typical compiler's built-in cosine function… so there's nothing to plot the data against.

Note that C's `double` type on most computers carries about 15 digits of precision. So for these algorithms, especially for the 20.2 and 23.1 digit versions, you'll need to use a data type that offers more bits. Some C's support a `long double`. But check the manual carefully! Microsoft's Visual C++, for instance, while it does support the `long double` keyword, converts all of these to `double`.

Accurate to about 14.7 decimal digits over the range $[0, \pi/2]$:

```
c1= 0.9999999999999806767
c2=-0.4999999999998996568
c3= 0.0416666666581174292
c4=-0.0013888888886113613522
c5= 0.0000248015828760424427
c6=-0.0000027556935768631581
c7= 0.00000000020858327958707
c8=-0.0000000000011080716368
cos(x)= c1 + x²(c2 + x²(c3 + x²(c4 + x²(c5 +
               x²(c6 + x²(c7 + x²*c8))))))
```

Accurate to about 20.2 decimal digits over the range $[0, \pi/2]$:

```
c1 = 0.9999999999999999999936329
c2 =-0.4999999999999999999948362843
c3 = 0.0416666666666666665975670054
c4 =-0.0013888888888888885302082298
c5 = 0.0000248015873014927464422297
c6 =-0.0000027557319209666748555
c7 = 0.00000000020876755667423458605
c8 =-0.0000000000114706701991777771
c9 = 0.00000000000000477687298095717
c10=-0.000000000000000015119893746887
cos(x)= c1 + x²(c2 + x²(c3 + x²(c4 + x²(c5 + x²(c6 +
               x²(c7 + x²(c8 + x²(c9 + x²*c10))))))))
```

Accurate to about 23.1 decimal digits over the range $[0, \pi/2]$:

```
c1 = 0.99999999999999999999999914771
c2 =-0.49999999999999999999991637437
c3 = 0.04166666666666666665319411988
c4 =-0.00138888888888888880310186415
c5 = 0.00002480158730158702330045157
c6 =-0.00000027557319223932256421489
c7 = 0.00000000208767569816541259159
c8 =-0.00000000001147074512677554323394
c9 = 0.00000000000004779454394066499917
c10=-0.000000000000000015612263428827781
c11= 0.0000000000000000000039912654507924
cos(x) = c1 + x²(c2 + x²(c3 + x²(c4 + x²(c5 + x²(c6 +
                x²(c7 + x²(c8 + x²(c9 + x²(c10 + x²*c11)))))))))
```

Tangent

The tangent of an angle is defined as `tan(x)=sin(x)/cos(x)`. Unhappily this is not the best choice, though, for doing an approximation. As cos(x) approaches zero the errors propagate rapidly. Further, at some points like $\pi/4$ (see the previous graphs of sine and cosine errors) the errors of sine and cosine reinforce each other; both are large and have the same sign.

So we're best off using a separate approximation for the tangent. All of the approximations we'll use generate a valid tangent for angles in the range of 0 to $\pi/4$ (0 to 45 degrees), so once again a range reduction function will translate angles to this set of values.

```
//
//  This is the main tangent approximation "driver"
// It reduces the input argument's range to [0, pi/4],
// and then calls the approximator.
// Enter with positive angles only.
//
// WARNING: We do not test for the tangent approaching infinity,
// which it will at x=pi/2 and x=3*pi/2. If this is a problem
// in your application, take appropriate action.
//
float tan_32(float x){
      int octant;                               // what octant are we in?

      x=fmod(x, twopi);                          // Get rid of values >2 *pi
      octant=int(x/qtrpi);              // Get octant # (0 to 7)
      switch (octant){
      case 0: return      tan_32s(x              *four_over_pi);
      case 1: return   1.0/tan_32s((halfpi-x)    *four_over_pi);
      case 2: return - 1.0/tan_32s((x-halfpi)    *four_over_pi);
      case 3: return -     tan_32s((pi-x)         *four_over_pi);
      case 4: return       tan_32s((x-pi)         *four_over_pi);
      case 5: return   1.0/tan_32s((threehalfpi-x) *four_over_pi);
      case 6: return - 1.0/tan_32s((x-threehalfpi) *four_over_pi);
      case 7: return -     tan_32s((twopi-x)      *four_over_pi);
      }
}
```

The code above does the range reduction and then calls `tan_32`. When using the higher precision approximations substitute the appropriate function name for `tan_32`.

The reduction works much like that for cosine, except that it divides the circle into octants and proceeds from there. One quirk is that the argument is multiplied by $4/\pi$. This is because the approximations themselves actually solve for $\tan((\pi/4)x)$.

The listings that follow give the algorithms needed.

Remember that tan(90) and tan(270) both equal infinity. As the input argument gets close to 90 or 270 the value of the tangent skyrockets, as illustrated on the following error charts. *Never take a tangent close to 90 or 270 degrees!*

tan_32 percentage error

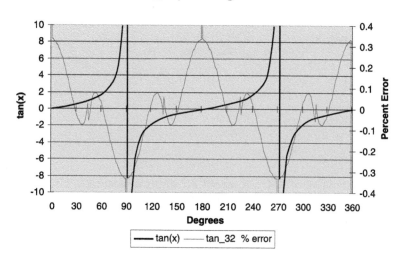

```
//  ************************************************************
//  ***
//  ***    Routines to compute tangent to 3.2 digits
//  ***    of accuracy.
//  ***
//  ************************************************************
//
//              tan_32s computes tan(pi*x/4)
//
//  Accurate to about 3.2 decimal digits over the range [0, pi/4].
//  The input argument is in radians. Note that the function
//  computes tan(pi*x/4), NOT tan(x); it's up to the range
//  reduction algorithm that calls this to scale things properly.
//
//  Algorithm:
//              tan(x)= x*c1/(c2 + x**2)
//
float tan_32s(float x)
{
const float c1=-3.6112171;
const float c2=-4.6133253;
float x2;                                 // The input argument squared

x2=x * x;
return (x*c1/(c2 + x2));
}
```

tan_32 computes the tangent of $\pi/4*x$ to about 3.2 digits of accuracy. Use the range reduction code to translate the argument to 0 to $\pi/4$, and of course to compensate for the peculiar "$\pi/4$" bias required by this routine. Note that the graphed errors are percentage error, not absolute.

tan_56 percentage error

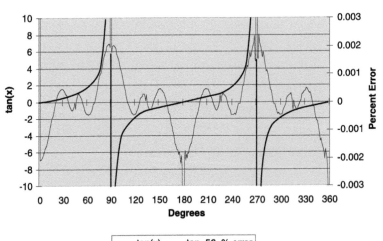

```
// ************************************************************
// ***
// ***   Routines to compute tangent to 5.6 digits
// ***   of accuracy.
// ***
// ************************************************************
//
//             tan_56s computes tan(pi*x/4)
//
//   Accurate to about 5.6 decimal digits over the range [0, pi/4].
//   The input argument is in radians. Note that the function
//   computes tan(pi*x/4), NOT tan(x); it's up to the range
//   reduction algorithm that calls this to scale things properly.
//
//   Algorithm:
//             tan(x)= x(c1 + c2*x**2)/(c3 + x**2)
//
float tan_56s(float x)
{
const float c1=-3.16783027;
const float c2= 0.134516124;
const float c3=-4.033321984;
float x2;                                  // The input argument squared

x2=x * x;
return (x*(c1 + c2 * x2)/(c3 + x2));
}
```

tan_56 computes the tangent of $\pi/4*x$ to about 5.6 digits of accuracy. Use the range reduction code to translate the argument to 0 to $\pi/4$, and of course to compensate for the peculiar "$\pi/4$" bias required by this routine. Note that the graphed errors are percentage error, not absolute.

tan_82 percentage error

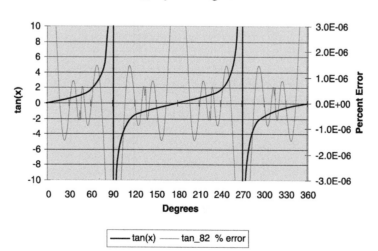

```
// ***********************************************************
// ***
// ***    Routines to compute tangent to 8.2 digits
// ***    of accuracy.
// ***
// ***********************************************************
//
//              tan_82s computes tan(pi*x/4)
//
//  Accurate to about 8.2 decimal digits over the range [0, pi/4].
//  The input argument is in radians. Note that the function
//  computes tan(pi*x/4), NOT tan(x); it's up to the range
//  reduction algorithm that calls this to scale things properly.
//
//  Algorithm:
//              tan(x)= x(c1 + c2*x**2)/(c3 + c4*x**2 + x**4)
//
double tan_82s(double x)
{
const double c1= 211.849369664121;
const double c2=- 12.5288887278448 ;
const double c3= 269.7350131214121;
const double c4=- 71.4145309347748;
double x2;                                  // The input argument squared

x2=x * x;
return (x*(c1 + c2 * x2)/(c3 + x2*(c4 + x2)));
}
```

tan_82 computes the tangent of $\pi/4*x$ to about 8.2 digits of accuracy. Use the range reduction code to translate the argument to 0 to $\pi/4$, and of course to compensate for the peculiar "$\pi/4$" bias required by this routine. Note that variables are declared as "double". The graphed errors are percentage error, not absolute.

tan_14 percentage error

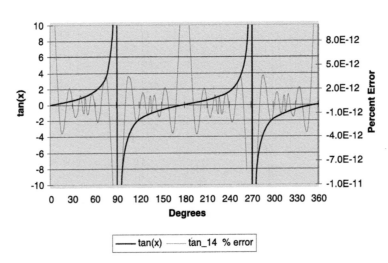

```
// ************************************************************
// ***
// ***   Routines to compute tangent to 14 digits
// ***   of accuracy.
// ***
// ************************************************************
//
//          tan_14s computes tan(pi*x/4)
//
//   Accurate to about 14 decimal digits over the range [0, pi/4].
//   The input argument is in radians. Note that the function
//   computes tan(pi*x/4), NOT tan(x); it's up to the range
//   reduction algorithm that calls this to scale things properly.
//
//   Algorithm:
//          tan(x)= x(c1 + c2*x**2 + c3*x**4)/(c4 + c5*x**2 + c6*x**4 + x**6)
//
double tan_14s(double x)
{
const double c1= -34287.4662577359568109624;
const double c2=   2566.7175462315050423295;
const double c3= -   26.5366371951731325438;
const double c4= -43656.1579281292375769579;
const double c5=  12244.4839556747426927793;
const double c6= -  336.611376245464339493;
double x2;                          // The input argument squared

x2=x * x;
return (x*(c1 + x2*(c2 + x2*c3)))/(c4 + x2*(c5 + x2*(c6 + x2))));
}
```

tan_141 computes the tangent of $\pi/4*x$ to about 14.1 digits of accuracy. Use the range reduction code to translate the argument to 0 to $\pi/4$, and of course to compensate for the peculiar "$\pi/4$" bias required by this routine. Note that variables are declared as "double". The graphed errors are percentage error, not absolute.

Higher Precision Tangents

Given a large enough polynomial there's no limit to the possible accuracy. A few more algorithms are listed here. These are all valid for the range of 0 to $\pi/4$, and all should use the previous range reduction algorithm to change any angle into one within this range. All take an input argument in radians, though it is expected to be mangled by the $\pi/4$ factor. The prior range reducer will correct for this.

No graphs are included because these exceed the accuracy of the typical compiler's built-in cosine function… so there's nothing to plot the data against.

Note that C's double type on most computers carries about 15 digits of precision. So for these algorithms, especially for the 20.2 and 23.1 digit versions, you'll need to use a data type that offers more bits. Some C's support a long double. But check the manual carefully! Microsoft's Visual C++, for instance, while it does support the long double keyword, converts all of these to double.

Accurate to about 20.3 digits over the range of 0 to $\pi/4$:

```
c1= 10881241.46289544215469695742
c2=-  895306.08705641459574470875755
c2=    14181.99563014366386894487566
c3=-      45.63638305432707847378129653
c4= 13854426.92637036839270054048
c5=- 3988641.46816307730070133878784
c6=    135299.47445500236808675591955
c7=-    1014.19757617656429288596025
tan(xπ/4)=x(c1 + x²(c2 + x²(c3 + x²*c4)))
              /(c5 + x²(c6 + x2(c7 + x²)))
```

Accurate to about 23.6 digits over the range of 0 to $\pi/4$:

```
c1=    4130240.55899602401344014622677
c2=-  349781.85625173816166310124877
c3=      6170.31775814249424533194434488
c4=-       27.94920941480194872760036319
c5=         0.017514380704038360266655630588
c6= 5258785.647179987798541780825
c7=- 1526650.54907294068677625989933
c8=     54962.516160629053611522305566
c9=-      497.49546028091726502450693737
tan(xπ/4)=x(c1 + x²(c2 + x²(c3 + x²(c4 + x²*c5))))
              /(c6 + x²(c7 + x²(c8 + x²(c9 + x²))))
```

Arctangent, Arcsine and Arccosine

The arctangent is the same as the inverse tangent, so arctan(tan(x))=x. It's often denoted as "atan(x)" or "tan⁻¹(x)".

In practice the approximations for inverse sine and cosine aren't too useful; mostly we derive these from the arctangent as follows:

```
Arcsine(x)      = atan(x/√(1-x²))

Arccosine(x)    = π/2 - arcsine(x)

                = π/2 - atan(x/√(1-x²))
```

The approximations are valid for the range of 0 to $\pi/12$. The following code, based on that by Jack Crenshaw in his *Math Toolkit for Real-Time Programming*, reduces the range appropriately:

```
//
//   This is the main arctangent approximation "driver"
// It reduces the input argument's range to [0, pi/12],
// and then calls the approximator.
//
//
double atan_66(double x){
double y;                               // return from atan__s function
int complement= FALSE;                  // true if arg was >1
int region= FALSE;                      // true depending on region arg is in
int sign= FALSE;                        // true if arg was < 0

if (x <0 ){
    x=-x;
    sign=TRUE;                          // arctan(-x)=-arctan(x)
}
if (x > 1.0){
    x=1.0/x;                            // keep arg between 0 and 1
    complement=TRUE;
}
if (x > tantwelfthpi){
    x = (x-tansixthpi)/(1+tansixthpi*x); // reduce arg to under tan(pi/12)
    region=TRUE;
}

y=atan_66s(x);                          // run the approximation
if (region) y+=sixthpi;                 // correct for region we're in
if (complement)y=halfpi-y;              // correct for 1/x if we did that
if (sign)y=-y;                          // correct for negative arg
return (y);

}
```

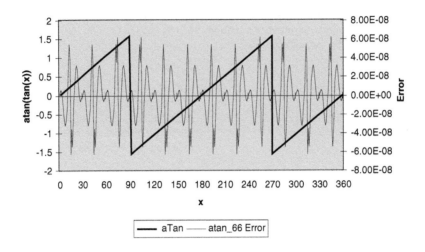

atan_66 Errors

aTan ——— atan_66 Error

```
// ************************************************************
// ***
// ***    Routines to compute arctangent to 6.6 digits
// ***    of accuracy.
// ***
// ************************************************************
//
//              atan_66s computes atan(x)
//
//   Accurate to about 6.6 decimal digits over the range [0, pi/12].
//
//   Algorithm:
//              atan(x)= x(c1 + c2*x**2)/(c3 + x**2)
//
double atan_66s(double x)
{
const double c1=1.6867629106;
const double c2=0.4378497304;
const double c3=1.6867633134;


double x2;                              // The input argument squared

x2=x * x;
return (x*(c1 + x2*c2)/(c3 + x2));
}
```

atan_66 computes the arctangent to about 6.6 decimal digits of accuracy using a simple rational polynomial. Its input range is 0 to $\pi/12$; use the previous range reduction code.

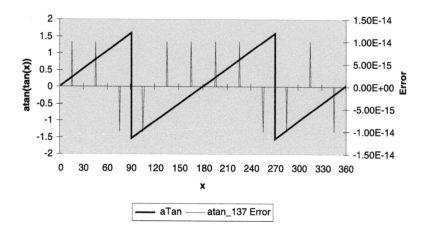

atan_137 Errors

```
//  ************************************************************
//  ***
//  ***   Routines to compute arctangent to 13.7 digits
//  ***   of accuracy.
//  ***
//  ************************************************************
//
//             atan_137s computes atan(x)
//
//   Accurate to about 13.7 decimal digits over the range [0, pi/12].
//
//   Algorithm:
//           atan(x)= x(c1 + c2*x**2 + c3*x**4)/(c4 + c5*x**2 + c6*x**4 + x**6)
//
double atan_137s(double x)
{
const double c1=  48.70107004404898384;
const double c2=  49.5326263772254345;
const double c3=   9.40604244231624;
const double c4=  48.70107004404996166;
const double c5=  65.7663163908956299;
const double c6=  21.587934067020262;

double x2;                               // The input argument squared

x2=x * x;
return (x*(c1 + x2*(c2 + x2*c3))/(c4 + x2*(c5 + x2*(c6 + x2))));
}
```

atan_137 computes the arctangent to about 13.7 decimal digits of accuracy using a simple rational polynomial. Its input range is 0 to $\pi/12$; use the previous range reduction code.

Math Functions

Jack Ganssle

It's not always possible to count on a runtime library to provide all of the resources needed to do various forms of math in an embedded system. The library might not include a needed function, or the library implementation might be slow or large. It's useful, therefore, to have a set of alternative formulations.

Gray Code

To convert the 8-bit Gray code number G to a binary number B:

```
BEGIN: set B0 through B7 = 1
B7     =  G7
IF B6  =  G5 THEN B5 = 0
IF B5  =  G4 THEN B4 = 0
IF B4  =  G3 THEN B3 = 0
IF B3  =  G2 THEN B2 = 0
IF B2  =  G1 THEN B1 = 0
IF B1  =  G0 THEN B0 = 0
```

To convert the binary number B to its Gray code representation G:

```
G = B xor (B >>2)
```

Integer Multiplication by a Constant

Many CPUs don't have a decent multiply instruction. When speed is important try reformulating the equation.

For instance, we know that multiplying by 2 is wasteful—it's far faster to do a shift. Why not use the same trick for multiplications by other constants? Here's a few; it's trivial to extend the idea for other constants:

```
A * 3  =  A * 2 + A
A * 5  =  A * 4 + A
A * 10 =  (A * 4 + A) * 2
```

Computing an Exclusive Or

The XOR of A and B can be found by:

```
A xor B = A + B - 2 (A and B)
```

Integer Square Roots

One approach is to sum all of the odd numbers from 1, till the sum is less than or equal to the number whose root you're looking for. Count the number of summations, which will be the square root. For instance, to compute the square root of 31:

    1 + 3 + 5 + 7 + 9 = 25; one more would be 36 which exceeds 31.

Another approach to compute a square root of a 32-bit number, returning a 16-bit result quickly (no more than 16 iterations), is:

```
typedef unsigned int       UInt16;
typedef unsigned long int UInt32;

static UInt16 isqrt (UInt32 x)
{
   UInt16 bit  = 16;

   UInt16 mask = 0x8000;
   UInt16 root = 0x0000;

   do
      {
         UInt32 acc = root | mask;

         if (acc * acc <= x)
            root |= mask;

         mask >>= 1;
      }
   while (—bit);

   return root;
```

Important Basic Math Operations

There are a number of basic relations between functions that can simplify your code.

```
arcsin(x)   = arctan(x/sqr(-x² + 2))
arccos(x)   = -arctan(x/sqr(-x²+1))+ pi/2
arcsec(x)   = arctan(x/sqr(x²-1))
arccsc(x)   = arctan(x/sqr(x²-1))  + (sgn(x)-1) * pi/2
arccot(x)   = arctan(x) + pi/2
sinh(x)     = (exp(x) - exp(-x))/2
cosh(x)     = (exp(x) + exp(-x))/2
tanh(x)     = exp(-x)/(exp(x) + exp(-x))*2 + 1
arcsinh(x) = log(x + sqr(x²+1))
arccosh(x) = log(x + sqr(x²-1))
arctanh(x) = log(1 + x)(1 - x))/2
```

IEEE 754 Floating Point Numbers

Jack Ganssle

IEEE 754 floating point is the most common representation today for real numbers on computers.

Floating-point represents reals using a base number and an exponent. For example, 123.456 could be represented as 1.23456×10^2. In hexadecimal, the number 123.abc might be represented as $1.23abc \times 16^2$.

IEEE floating point numbers have three components: the sign, the exponent (also known as the characteristic), and the mantissa. The mantissa is composed of the *fraction* and an implicit leading digit (explained below). The exponent is always applied to an implied 2, so is really $2^{exponent}$.

The numbers are organized in memory as follows:

	Sign	Exponent	Fraction	Bias
Single Precision (32 bits)	1 bit, msb (bit 31)	8 bits in positions 30-23	23 bits in positions 22-00	127
Double Precision (64 bits)	1 bit, msb (bit 63)	11 bits in positions 62-52	52 bits in positions 51-00	1023

The sign bit is a 0 to denote a positive number; 1 is a negative number.

To support both negative and positive exponents a *bias* is added to the actual—127 for single-precision floats. If the actual exponent (the one used in doing the math) is zero, 127 is stored in the exponent field. A stored value of 200 indicates an exponent of (200–127), or 73. For reasons discussed later, exponents of –127 (all 0s) and +128 (all 1s) are reserved for special numbers.

Double precision numbers use a bias of 1023, in an 11 bit field, so larger and smaller numbers can be represented.

The *mantissa* represents the number itself, without exponent. It is composed of an implied leading bit and the fraction bits.

The value of a single precision float is:

$$(-1)^S \, 2^{\,E\text{-}127} \, (1.M)$$

That of a double is:

$$(-1)^S \, 2^{E\text{-}1023} \, (1.M)$$

Where:

S is the sign

E is the exponent stored in memory

M is the mantissa

Special values

The IEEE-754 standard supports some special values:

Name	Exponent	Fraction	sign	Exp Bits	Mantissa Bits
+0	min−1	= 0	+	All zeros	All Zeros
−0	min−1	= 0	−	All zeros	All Zeros
Number	min ≤ e ≤ max	any	any	Any	Any
+∞	max+1	= 0	+	All ones	All zeros
−∞	max+1	= 0	−	All ones	All zeros
NaN	max+1	≠ 0	any	All ones	Any

Zero is represented as all zeroes in the mantissa and exponent, with a zero or one in the sign bit. This, plus and minus zeroes are both legal. The standard does require comparisons using +0 and −0 to return the same result.

Some computations (like divide by zero or $\sqrt{[(-1)]}$) generate undefined results—defined as Not a Number (NaN).

Whenever a NaN participates in any operation the result is NaN. All comparison operators $(=, <, \le, >, \ge)$ except not-equals should return false when NaN is one of the operands.

Sources of NaN:

Operation	Produced by
+	∞+(−∞)
×	0×∞
/	0/0, ∞/∞
REM	x REM 0, ∞REM y
√[]	√x (when x < 0)

Operations with infinity:

Operation	Result
n / ±Infinity	0
±Infinity x ±Infinity	±Infinity
±nonzero / 0	±Infinity
Infinity + Infinity	Infinity
±0 / ±0	*NaN*
Infinity - Infinity	*NaN*
±Infinity / ±Infinity	*NaN*
±Infinity x 0	*NaN*

Parameter	Single	Double
p (precision, in bits)	24	53
Decimal digits of precision	7.22	15.95
Mantissa's MS-Bit	hidden bit	hidden bit
Actual mantissa width in bits	23	52
E_{max}	+127	+1023
E_{min}	−126	−1022
Exponent *bias*	+127	+1023
Exponent width in bits	8	11
Sign width in bits	1	1
Format width in bits	32	64
Range Magnitude Maximum $2^{E_{max}+1}$	3.4028E+38	1.7976E+308
Range Magnitude Minimum $2^{E_{min}}$	1.1754E−38	2.2250E−308

SECTION

IV

Real-Time

Introduction

Jack Ganssle

If you study the literature of software engineering you'll find an interesting pattern: academics often speak of "embedded" (when they refer to it at all) and "real-time" as two different arenas. Astonishingly, Barry Boehm, in his landmark creation of the COCOMO estimation model considered airline reservation systems, which must give an answer pretty quickly, as "real time."

Yet the embedded world is usually a real-time world. When an interrupt comes, you usually have little time to service it. An external device needs attention… *now*.

In this section you'll find hints on selecting a real-time operating system, written by Jean Labrosse, probably the best-known authority on RTOSes. His uC/OS has been included in hundreds of products… and is safety certified to the rigid DO-178B standard. It's correct… and it's beautiful. Check out his book and examine the RTOS's source code – it's stunning. Well structured, perfectly commented, conforming to a rigid standard.

Do beauty and reliability go hand in hand? Empirical evidence suggests so.

Reentrancy and latency are important and overlooked concepts that I've addressed in two chapters here. Another worthwhile source is *An Embedded Software Primer* by David Simon, a well-written introduction to some of these issues.

Jakob Engblom and Sandeep Ahluwalia have provided chapters on optimizing your code. Jakob's experience working at a compiler company gives him insights into working with the compiler, rather than at odds with it. Sandeep gives his optimization observations from a user perspective.

Above all, when creating your real-time code, remember to think in the time domain as well as in the procedural domain. Time is a precious commodity, as valuable as ROM and RAM. We have but a single second, each second, to get our time-critical actions accomplished. Slip by even a microsecond, and your user may experience catastrophe.

Real-Time Kernels

Jean J. Labrosse

Jean J. Labrosse is President of Micrium, Inc., a provider of high quality embedded software solutions. He has a Master's degree in electrical engineering from the University of Sherbrooke in Sherbrooke, Canada, and has been designing embedded systems for close to 20 years. Mr. Labrosse is also the author of two books in the field: *MicroC/OS-II, The Real-Time Kernel* (ISBN 1-57820-103-9) and *Embedded Systems Building Blocks, Complete and Ready-to-Use Modules in C* (ISBN 0-87930-604-1). Mr. Labrosse has written many articles for magazines and is a regular speaker at the Embedded Systems Conference and serves on the Advisory Board for the conference.

Introduction

In this chapter, I will describe what a real-time kernel is and provide some information about how to use one. I will then provide some selection criteria in case you are faced with the task of selecting a real-time kernel.

I will be using an actual kernel called µC/OS-II in the examples. µC/OS-II is a highly portable, ROMable, scalable, preemptive, real-time, multitasking kernel and was specifically designed for embedded applications. µC/OS-II is written in ANSI C and the inner workings of µC/OS-II are described in the book *MicroC/OS-II, The Real-Time Kernel* (ISBN 1-57820-103-9) written by Jean J. Labrosse. µC/OS-II was certified in multiple avionics products by the Federal Aviation Administration (FAA) for use in commercial aircraft by meeting the demanding requirements of the RTCA DO-178B standard for software used in avionics equipment. The book includes a CD that contains ALL the source code for µC/OS-II. Since its introduction back in 1992, µC/OS has been used in hundreds of products. There are over 100 ports for µC/OS-II that you can download for FREE from the Micrium web site: www.Micrium.com.

What is a Real-Time Kernel?

A real-time kernel (from now on, a *Kernel*) is software that manages the time of your microprocessor to ensure that time-critical events are processed as efficiently as possible. The use of a kernel simplifies the design of your product because it allows your system to be divided into multiple independent elements called *tasks*.

What is a task?

A task is a simple program that thinks it has the microprocessor all to itself. Each task has its own stack space and each task is dedicated to a specific function in your product. The kernel is responsible for keeping track of the top-of-stack for each of the different tasks. When you design a product with a kernel, you assign each task a priority based on what you believe the importance of the task is. The design of a real-time system now consists of splitting the work to be done into tasks which are responsible for a portion of the problem. For example, each of the following functions can be handled by one or more tasks:

- Reading sensors (temperatures, pressures, switches ...)
- Computations
- Control loops
- Updating outputs (actuators, relays, lights ...)
- Monitoring
- Protocol stacks
- Keyboard scanning
- Operator interface
- Display (character-based or graphics)
- Time-of-day clock/calendar
- Serial communications
- Data logging
- Etc.

A task is an infinite loop and looks as shown below.

```
void MyTask (void)
{
    while (1) {
        // Do something (Your code)
        // Wait for an event, either:
        //    Delay for 'n' clock ticks
        //    Wait on a semaphore
        //    Wait for a message
        //    Other
        // Do something (Your code)
    }
}
```

You, as the application programmer, need to provide information to the kernel about your tasks. This is done by calling a service provided by the kernel to *create a task*. Specifically, you need to tell the kernel the address of your task (i.e., the address of MyTask() in the code above), the priority you want to give to this task and, how much stack space you believe the task will need. Generally speaking, you determine the priorities of your tasks based on their importance. When you create a task, the kernel maintains a data structure for each task called a *Task Control Block (TCB)*. The TCB contains the task priority, the state of the task (ready or waiting), the current top-of-stack for the task and other information about each task.

In the task body, within the infinite loop, you need to include calls to kernel functions so that your task waits for some *event*. One task can tell the kernel "Run me every 100 milliseconds," another task can say "I want to wait for a packet to arrive from an Ethernet controller," yet another task can say "I want to wait for an analog-to-digital conversion to complete," and so on. Events are generally generated by your application code, either from ISRs (Interrupt Service Routines) or issued by other tasks. In other words, an ISR can send a *message* to a task to *wake up* the task. Similarly, a task can send a message to another task. This event mechanism is the kernel's way to ensure that each task doesn't consume all of the CPU's time. In other words, if you don't tell the kernel to wait for these events, a task would be a true infinite loop and thus consume all of the CPU's time.

Most kernels are *preemptive*. This means that the kernel will always try to execute the highest priority task that is ready to run. The execution profile of a system designed using a *preemptive* kernel is shown in Figure 15-1.

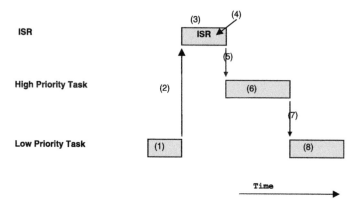

Figure 15-1: Execution profile of a preemptive kernel.

F1-(1)   A low priority task is executing.

F1-(2)   The *event* that a high priority task was waiting for occurs and causes an interrupt on the microprocessor.

F1-(3)   The microprocessor saves the *context* of the task that was running onto that task's stack and vectors to the ISR that services the event. The context of a task generally consists of the CPU registers (more on this later).

F1-(4)   The ISR calls a service provided by the kernel to wake up the high priority task and thus makes it ready for execution.

F1-(5)   Upon completion, the ISR invokes the kernel once more and the kernel notices that a more important task (than the one that was running prior to the interrupt) is ready to run. The kernel will thus not return to the interrupted task but instead, run the more important task.

F1-(6)   The higher priority task executes to completion or, until it desires to wait for another event.

F1-(7)   When the high priority task calls a function to wait for an event, the kernel *suspends* the high priority task and *resumes* the lower priority task, exactly where it was interrupted.

F1-(8)   The lower priority task continues to execute.

If a task that is currently running makes a more important task ready to run, then the current task is suspended and the more important task gets to run. Figure 15-2 shows how the kernel can context switch between multiple tasks.

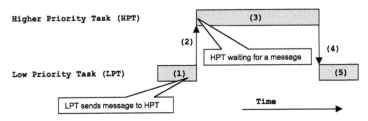

Figure 15-2: Context switching between multiple tasks.

F2-(1)   A low priority task is executing. The low priority task (LPT) sends a message to a high priority task (HPT).

F2-(2)   Because the higher priority task is more important than the low priority task, the kernel *suspends* execution (saves the context) of the low priority task and *resumes* execution (restores the context) of the higher priority task. It is assumed that the HPT was waiting for a message. The two tasks obviously need to agree about the kind of messages being sent.

F2-(3)   The higher priority task executes to completion (unless it gets interrupted).

F2-(4)   The higher priority task will want to wait for another message and thus the kernel will suspend the task and resume the low priority task.

F2-(5)   The low priority task continues execution.

If all the tasks in a system are waiting for their 'events' to occur then the kernel executes an internal task called the *Idle Task* which is basically an empty infinite loop. The Idle Task never waits for any event and thus, the CPU *spins* in the infinite loop until events occur. If you design a product that runs on batteries and the processor you are using offers low power *sleep modes* then you could place the CPU in one of those modes when in the idle task. In other words, the idle task code would look as follows:

```
void OSTaskIdle (void)
{
    while (1) {
        Place CPU in LOW POWER sleep mode;
    }
}
```

When an interrupt occurs (i.e., an event that one of the tasks is waiting for occurs), the CPU is taken out of sleep mode and the ISR is serviced. The ISR will most likely make a task ready-to-run and the kernel will thus 'switch' out of the idle task and then, execute the task that services the event. When the task is done servicing the event, and there are no other tasks to run, the kernel will 'switch' the CPU back to the idle task and run the code immediately following the instruction within the infinite loop (i.e. branch back to the top of the loop).

The Clock Tick

Just about every kernel needs a time source to keep track of delays and timeouts. This time source is called a *clock tick* and generally comes from a hardware timer that interrupts the CPU every 10 to 100 milliseconds. In fact, it's quite common to use the power line frequency of your product as the tick source and thus obtain either 50 Hz (Europe) or 60 Hz (North America). When the clock tick interrupt occurs, the kernel is notified and the kernel determines if any of the tasks waiting for time to expire need to be made ready-to-run. In other words, if a task needs to suspend execution for a period of 10 ticks then, the task will be made ready-to-run on the 10th clock tick interrupt. At that point, the kernel will decide whether the readied task is now the most important task and, if it is, the kernel will resume execution of the new task and will simply suspend execution of the task that was interrupted by the clock tick interrupt. The code below shows a task delaying itself for 10 ticks by calling the μC/OS-II function OSTimeDly().

```
void MyTask (void)
{
    while (1) {
        OSTimeDly(10);    // Suspend current task for 10 ticks!
        // Task code (i.e. YOUR code)
    }
}
```

As mentioned above, the clock tick interrupt is used to ready tasks that are waiting for events to occur in case these events do not occur within a user specified timeout. Specifically, the task below expects a message to be sent to it by either another task or an ISR. OSQPend() is another µC/OS-II function that allows your tasks to receive messages and contains three arguments. The first argument is a pointer to an object called a message queue which is used to hold the actual messages being sent, the second argument is the amount of time you want the task to wait for a message and is specified in clock ticks. The last argument is a pointer to an error code that the kernel returns to indicate the outcome of the OSQPend() call. Upon return from the OSQPend() call, your code simply needs to examine the error code to determine whether your task did in fact receive a message or whether your task was readied because it timed out waiting.

```
void MyTask (void)
{
    void    *pmsg;
    OS_ERR   err;

    while (1) {
        pmsg = OSQPend(MyMsgQ, 10, &err);
        if (err == OS_ERR_TIMEOUT) {
            // Message did not arrive within 10 ticks
        } else {
            // Message received, process message pointed to by 'pmsg'.
        }
    }
}
```

Scheduling

Scheduling is a function performed by the kernel to determine whether a more important task needs to run. Scheduling occurs under one of the following conditions:

1. An ISR (Interrupt Service Routine) sends a message to a task by using a *service* provided by the kernel. If the task receiving the message is more important than the task that was interrupted then, at the end of all nested ISRs, the task receiving the message will resume execution and the interrupted task will continue to be suspended (see Figure 15-3A). The outcome of scheduling in this case is a context switch (described later in details). If the task receiving the message is less important than the task that was interrupted then, at the end of all nested ISRs, the interrupted task will be resumed (see Figure 15-3B). In other words, there will be no context switch.

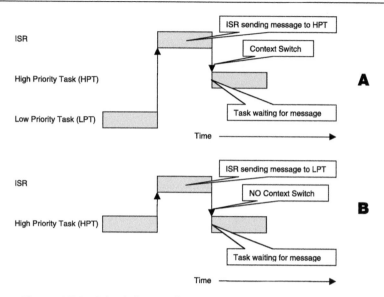

Figure 15-3: Scheduling and Context Switching because of ISR.

2. A task sends a message to another task using a service provided by the kernel. If the receiving task is more important than the sending task then the kernel will switch to the receiving task (see Figure 15-4) otherwise, the sending task continues execution.

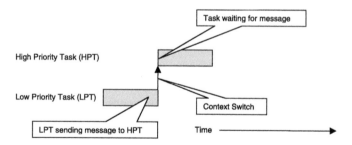

Figure 15-4: Scheduling and Context Switching because of Task.

3. A task desires to wait for time to expire (e.g. by calling OSTimeDly()). In this case, this task is suspended and the kernel selects and runs the next most important task that is ready-to-run.

In order for the kernel to run a task, the task must be *Ready-To-Run*. In other words, if a task was waiting for an event and the event occurs, the kernel makes the task eligible to run. The scheduler will only run tasks that are ready.

Context Switching

As previously mentioned, the context of a task generally consists of the contents of all the CPU registers. To suspend execution of a task, the kernel simply needs to save the CPU registers. Similarly, to resume execution of a task, the kernel simply needs to restore the CPU registers from a saved context. Figure 15-5 shows a context switch for a Motorola 68HC11 CPU.

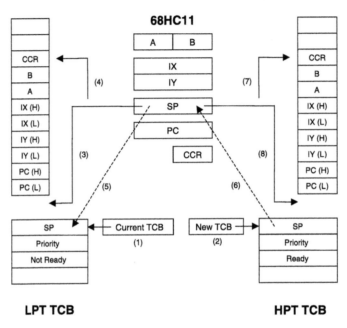

Figure 15-5: Context Switching on a Motorola 68HC11 CPU.

F5-(1) The kernel maintains a pointer to the current task's TCB (Task Control Block).

F5(2) When the scheduler decides to switch to another task, it updates a pointer to the TCB of the new task.

F5(3) The stack pointer of the CPU points to the current task's stack.

F5(4) To perform a context switch, the kernel starts by first saving all the CPU registers onto the current task's stack. For the 68HC11 processor, the kernel can simply execute a SWI (Software Interrupt) instruction. The SWI instruction pushes all of the CPU registers as shown.

F5(5) The current value of the SP (stack pointer) register is then saved by the kernel into the task's TCB.

F5(6) The kernel then loads the SP register with the new task's top-of-stack pointer from the new task's TCB.

F5(7) The SP register then points to the top-of-stack of the new task.

F5(8)   The kernel then executes a RTI instruction which causes the CPU to automatically pop all of the CPU registers off the stack. At this point, the CPU is running code in the new task.

The process is summarized in the following pseudo-code:

Scheduling and Context Switch on 68HC11 CPU:

```
{

    Find the HPT that's ready-to-run (point to its TCB);
    Save the CPU registers onto the current task's stack;
    Save the SP register into the current task's TCB;
    Load the SP register with the new task's top-of-stack from its TCB;
    Execute a return from interrupt instruction;

}
```

On the 68HC11 CPU, a context switch for µC/OS-II is less than 10 lines of assembly language code!

Kernel Services

At this point, you have learned that:

1.   A kernel allows you to split your application into multiple tasks,

2.   A kernel requires a time source (the clock tick) to allow for tasks to suspend themselves until a certain amount of time expires or to provide timeouts when receiving messages.

3.   A task or an ISR can send messages to other tasks.

4.   A preemptive kernel always tries to run the most important task that is ready-to-run. The scheduler is invoked whenever an event occurs and there is the possibility of having a more important task that is ready-to-run and thus needs to get the CPU's attention.

5.   A context switch is used to save the current state of a task and resume a more important task.

Most commercial kernels provide many more services for you to use. We will only cover a few of the basic ones in this chapter.

Kernel Services, Semaphores

One of the most basic services is to ensure exclusive access to common resources. Specifically, kernels provide a mechanism called a *semaphore* that your tasks need to use in order to access a shared variable, array, data structure, I/O device, etc. The code below shows two tasks that need to access (read and or modify) data.

```
OS_EVENT   *MyDataSem;

void   Task1 (void)
{
    INT8U   err;

    while (1) {
        OSSemPend(MyDataSem, 0, &err);
        // Access the shared data!
        OSSemPost(MyDataSem);
        :

        :

    }
}

void   Task2 (void)
{
    INT8U   err;

    while (1) {
        OSSemPend(MyDataSem, 0, &err);
        // Access the shared data!
        OSSemPost(MyDataSem);
        :

        :

    }
}
```

Before the data can be accessed, each task must 'obtain' the semaphore. For μC/OS-II, this is accomplished by calling a function called OSSemPend(). If the semaphore is 'available', the call to OSSemPend() returns and the task can access the resource. If another task 'has' the semaphore then the task that wants it will be 'blocked' and placed in a list waiting for the semaphore to be 'released'. When the task that owns the semaphore is done with accessing the resource, it calls OSSemPost() which releases the semaphore. At that point, the kernel determines whether there are any task waiting for the semaphore and if there is, the kernel 'passes' the semaphore to the highest priority task waiting for the semaphore. The kernel then determines whether the readied task is more important than the task that released the semaphore and, if it is, the kernel context switches to the more important task. Conversely, if the task that released the semaphore is more important than the task that was readied then, the kernel returns to the task that released the semaphore and executes the code immediately following the OSSemPost() call.

There are three types of semaphores: binary, counting and mutexes. A *binary semaphore* is used to access a single resource. For example, you can use a binary semaphore to access a display. Here I assume that your tasks are sharing a single display device (CRT, LCD, etc.). A *counting semaphore* is used when you need to access any one of multiple identical resources. For example, you would use a counting semaphore to access a buffer from a pool of identical buffers. The value of the semaphore would be initialized to the number of buffers in that pool. Once all buffers are allocated to tasks then further requests for buffers would have to be blocked (by the kernel) until buffers are released. The last type of semaphore is called a *Mutual Exclusion Semaphore* or, *Mutex* for short. A mutex is a special type of binary semaphore that is also used to access single resources. The difference between this type of semaphore and a regular binary semaphore is that the mutex has 'intelligence' built into it. Specifically, a mutex is used to reduce *priority inversions* which are inherent to the other types of semaphores.

Priority inversion is a problem in real-time systems and occurs mostly when you use a real-time kernel. Figure 15-6 illustrates a priority inversion scenario. Task 1 has a higher priority than Task 2, which in turn has a higher priority than Task 3.

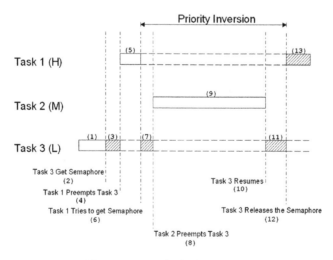

Figure 15-6: Priority inversion.

F6(1)   Task 1 and Task 2 are both waiting for an event to occur and Task 3 is executing.

F6(2)   At some point, Task 3 acquires a semaphore, which it needs before it can access a shared resource.

F6(3)   Task 3 performs some operations on the acquired resource.

F6(4)   The event that Task 1 was waiting for occurs and thus, the kernel suspends Task 3 and starts executing Task 1 because Task 1 has a higher priority.

F6(5)
F6(6)   Task 1 executes for a while until it also wants to access the resource (i.e., it attempts to get the semaphore that Task 3 owns). Because Task 3 owns the resource, Task 1 is placed in a list of tasks waiting for the semaphore to be freed.

F6(7)
F6(8)   Task 3 is resumed and continues execution until it is preempted by Task 2 because the event that Task2 was waiting for occurred.

F6(9)
F6(10)   Task 2 handles the event it was waiting for and, when it's done, the kernel relinquishes the CPU back to Task 3.

F6(11)
F6(12)   Task 3 finishes working with the resource and releases the semaphore. At this point, the kernel knows that a higher priority task is waiting for the semaphore, and a context switch is done to resume Task 1.

F6(13)   At this point, Task 1 has the semaphore and can access the shared resource.

The priority of Task 1 has been virtually reduced to that of Task 3 because it was waiting for the resource that Task 3 owned. The situation was aggravated when Task 2 preempted Task 3, which further delayed the execution of Task 1.

A mutex corrects the problem by implementing a scheme called *priority inheritance*. Figure 15-7 illustrates what happens when a kernel such as μC/OS-II supports priority inheritance.

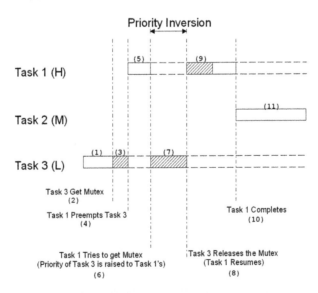

Figure 15-7: Priority Inheritance.

F7(1)

F7(2)   As with the previous example, Task 3 is running but this time, acquires a mutex to access a shared resource.

F7(3)

F7(4)   Task 3 accesses the resource and then is preempted by Task 1.

F7(5)

F7(6)   Task 1 executes and tries to obtain the mutex. The kernel sees that Task 3 has the mutex and knows that Task 3 has a lower priority than Task 1. In this case, the kernel raises the priority of Task 3 to the same level as Task 1.

F7(7)   The kernel places Task 1 in the mutex wait list and then resumes execution of Task 3 so that this task can continue with the resource.

F7(8)   When Task 3 is done with the resource, it releases the mutex. At this point, the kernel reduces the priority of Task 3 to its original value and looks in the mutex wait list to see if a task is waiting for the mutex. The kernel sees that Task 1 is waiting and gives it the mutex.

F7(9)   Task 1 is now free to access the resource.

F7(10)

F7(11)   When Task 1 is done executing, the medium-priority task (i.e., Task 2) gets the CPU. Note that Task 2 could have been ready to run any time between F7(3) and F7(10) without affecting the outcome. There is still some level of priority inversion that cannot be avoided but far less than in the previous scenario.

Kernel Services, Message Queues

A *message queue* (or simply a queue) is a kernel object that allows a task or an ISR to send messages to another task. Message queues are generally implemented as FIFO (First-In-First-Out) circular buffers. The number of messages that can be sent through a queue is generally configurable. In fact, an application can define multiple queues, each having its own storage capacity. You define the meaning of messages in your application. In other words, you define the meaning of messages and not the kernel; the kernel is just there to provide you with the message passing mechanism. With μC/OS-II, a message is simply a pointer and you define what the pointer points to.

Figure 15-8 shows an example of a task detecting an error condition and another task responsible for handling the condition.

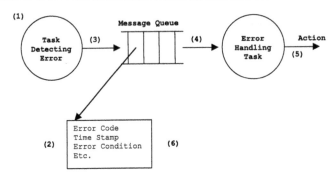

Figure 15-8: Message queues.

F8(1)   The task detects an error condition (temperature too high, oil pressure too low, etc.).

F8(2)   The task that detects the error then allocates a storage buffer, fills it with information about the error (error code, time stamp, condition that caused the error, etc.).

F8(3)   The task then invokes a service provided by the kernel to 'post' the message to the error handling task. With µC/OS-II, this is done using the OSQPost() function call. At this point, the kernel determines whether there is a task waiting for a message to arrive. If there is, the kernel makes the highest priority task that is waiting for the message ready-to-run and actually passes the message directly to that task. Although you would only have one task handle error messages, it's possible to have multiple tasks and thus, the kernel always ensures that the most important task waiting will actually receive the message. If the task that receives the message is more important than the task sending the message then the kernel will perform a context switch and thus, immediately run the more important task. If the receiving task is less important than the sending task, the OSQPost() function returns to the sender code and that task continues execution.

F8(4)   The receiving task (the error handler) handles the message sent when it becomes the most important task to run. To receive a message, this task needs to also make a call to a kernel provided function. For µC/OS-II, you'd call OSQPend(). When a message is received, the error handling task would process the message. Of course, both the sender and the receiver need to agree about the contents of the message being passed.

F8(5)   The error handling task could then 'log' the information received to a file (i.e. disk or flash), turn ON a light, update a display, energize a relay, etc.

F8(6)   Finally, when the task is done with the message, it would de-allocate the buffer so it can be used for another message.

Kernel Services, Memory Management

Your application can allocate and free dynamic memory using any ANSI C compiler's `malloc()` and `free()` functions, respectively. However, using `malloc()` and `free()` in an embedded real-time system is dangerous because, eventually, you may not be able to obtain a single contiguous memory area due to *fragmentation*. Fragmentation is the development of a large number of separate free areas (i.e., the total free memory is fragmented into small, non-contiguous pieces). Execution time of `malloc()` and `free()` are also generally nondeterministic because of the algorithms used to locate a contiguous block of free memory.

µC/OS-II and most kernels provide an alternative to `malloc()` and `free()` by allowing your application to obtain fixed-sized *memory blocks* from a *partition* made of a contiguous memory area, as illustrated in Figure 15-9. All memory blocks within a specific partition are the same size and the partition contains an integral number of blocks. The kernel maintains a list of free memory blocks as shown. Allocation and deallocation of these memory blocks is done in constant time and is deterministic.

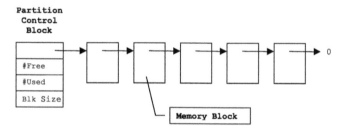

Figure 15-9: Memory partitions.

When a task requires a memory buffer, it simply asks the kernel to provide it with a memory block from one of its memory partitions. When a task is done with a buffer (i.e. memory block), it simply returns it to the proper memory partition by calling a service provided by the kernel.

Do You Need a Kernel?

A kernel allows your product to be more responsive to real-time events. Also, the responsiveness of high priority tasks is virtually unaffected by low priority tasks. Multitasking allows you to logically and cleanly split your application. Not all tasks need to be time critical and in fact, it's a common mistake to think that you don't need a kernel if you have few time critical tasks. As applications and products become more and more complex, a kernel allows you to better manage the tasks that need to be performed by your products.

A product using a 4-bit CPU will most likely not be able to use a kernel because these processors are typically used for very simple, low cost and dedicated applications. Products built using an 8-bit CPU could use the benefits of a kernel depending on the complexity of the product. I've designed many products using 8-bit CPUs and found that a kernel would have simplified my programming life without adding too much cost. In fact, as customers required additional features, I often regretted not taking a step back and adding a kernel to some of those products. If you use a 16-bit CPU, I'd highly recommend that you consider using a kernel and, if you use a 32-bit CPU, you really ought to use one.

Can You Use a Kernel?

This question is difficult to answer since it's based on a number of factors as described below:

CPU The CPU you use may determine whether you can use a kernel or not. Specifically, the kernel will consume some processing power when you make service calls. A kernel should consume only about 2 to 5% of the processing power. Also, because each task requires its own stack space, your processor needs to allow saving and changing the value of the stack pointer at run time. A processor like the Motorola 68HC05, for example, could not run a kernel because you cannot save and alter the SP register.

ROM The amount of memory available to hold your code (also known as code space) may determine whether you can use a kernel. Because a kernel is software that you add to your product, it requires code space. Depending on the kernel, you may need between 4 kbytes and 200 kbytes. Some kernels like µC/OS-II are scalable and you can thus reduce the 'footprint' based on the features you need. For example, on a Motorola 68HC11, µC/OS-II requires between 4 kbytes and 16 kbytes. Of course, you need to allocate sufficient code space for your own code.

RAM The amount of RAM (also known as data space) may determine whether you can use a kernel. Specifically, the kernel requires RAM to maintain TCBs, semaphores, message queues and a few other variables. More importantly, because each task requires its own stack space, you need to have sufficient RAM for the stack of all the tasks you intend to have. The amount of stack space for each task depends on the number of nested functions, the number of arguments passed, the number of local variables allocated within those functions, interrupt nesting, etc. You may find yourself needing between about 100 bytes to a few kbytes for certain tasks. Of course, you need to allocate sufficient RAM for your own application.

COST There are many elements to the cost equation. For one thing, since a kernel requires extra ROM and RAM, you might need to use slightly more expensive hardware in order to accommodate their requirements.

 Also, kernels are available commercially and the companies that sell them charge either using a royalty based model or a licensing model. Royalty based kernels simply mean that you must pay the kernel vendor for every unit that is shipped containing the kernel. They basically treat the kernel as if it was a chip. The price per unit varies depending on the number of units but typically ranges from pennies per unit to a few hundred dollars per unit.

A licensing model charges either on a per-product basis, on a product-line or on a per-CPU type used. Licensing costs vary greatly between kernel vendors but range from a few thousand dollars per product to tens of thousands of dollars. However, these are one-time fees and allow you to sell any number of units of the product you manufacture.

Another element to the cost equation is the learning curve. Like any piece of software you purchase, you must invest some time understanding how to use the kernel that you choose. You could easily spend between a week and a few months.

Selecting a Kernel?

There are currently close to 150 different commercial kernels to choose from. It's virtually impossible to evaluate every single one. Like most engineers, when I select a product, I build a matrix listing the features I'm looking for and of course, I have to locate products that will contain sufficient data about their product to allow me to fill in the matrix. Selecting a kernel should be no different. Here are some of the items to look for:

Popularity   Although not a technical issue, a kernel's popularity may be something for you to consider. Why? Because, you may readily find people that are already trained to use the kernel. Also, because others may have already gone through the selection process you will be going through, they may have selected the kernel for reasons similar to yours.

Reputation   The reputation of the kernel vendor is an important one to consider. Is the vendor known to provide a good product? Get references and call/contact them. Find out why they selected the kernel. How was technical support? How long did it take to get the kernel running? Does the company web site contain valuable information about the products?

Source Code   Some kernel vendors provide source code. Having the source code might give you some feeling of security in case the kernel vendor goes out of business or in case you need to alter the kernel to satisfy specific requirements. Is this important to you? If so, ask the kernel vendor to provide you with samples of the actual code. Does the code meet your quality expectations? Is the code well commented? Does the kernel vendor provide you with information about how the kernel works or will you have to figure that out for yourself?

Portability   Portability is the ability of a kernel to be used with different processors. This is an important feature for a kernel because it allows you to migrate your application code from one product to another even though the different products use different processors. Find out whether the kernel supports the processor or processors you are planning on using and if the processor is not supported whether it's difficult to port the kernel.

Scalability   Scalability is the ability of a kernel to adjust its footprint (ROM and RAM requirements) based on the features of the kernel that you require. In other words, can the amount of memory required by the kernel be reduced if you don't need all its features in a given product.

Preemptive   Of all the kernels that are commercially available, only a handful are 'non-preemptive'. If you don't believe you need a preemptive kernel then you just simplified your selection! However, there's a good likelihood that you will require a preemptive kernel.

#Tasks   Does the kernel support the number of tasks anticipated for your product? Because a kernel doesn't support thousands of tasks doesn't necessarily make it a bad choice. In fact, a kernel that limits the number of tasks may do so for performance reasons. Because each task requires its own stack space, the amount of RAM in your product may limit the number of tasks you can have anyway. Because you have a kernel doesn't mean you should create a task for everything!

Stack sizes   It's important that the kernel allows you to determine the size of each task on a task-per-task basis. In other words, a kernel should not force you to make the size of each task stack the same.

Services   Does the kernel provide the services you expect to need? Semaphores? Mutexes? Timers? Mailboxes? Queues? Memory management? Etc.

Performance   As I mentioned, a kernel should only require about 2 to 5% of the CPU's bandwidth. However, there are some aspects of the kernel that you might want to consider. One of the most important ones is the kernel interrupt disable time. In other words, the amount of time that the kernel will disable interrupts in order to manipulate critical sections of code. Another aspect is whether the kernel services are deterministic or not. In other words, are you able to determine how much time each service will take to execute and also, are these times constant or do they depend on the number of tasks in your system.

Performance   In as many projects as not, performance is a big issue, but understanding an RTOS's impact on your system isn't easy. When you compare vendors' benchmarks make sure you understand what they are measuring. What evaluation boards did each vendor use? What was the microprocessor's clock speed? What memory system was used? How many wait states were used for memory accesses? Only then can you make a fair comparison.

Certification   Certain industries have specific regulations or standards they demand from software. In some cases, off-the-shelf operating systems may not meet those needs (such as avionics, medical devices, automotive and safety systems). μC/OS-II has been certified in a number of medical and avionics applications and is suitable for safety critical systems. Only a handful of commercial kernels can claim this mark of robustness and quality.

Tools   Kernels are either provided in source or pre-compiled object code. If you get the source code, you ought to be able to compile the kernel with just about any compiler/assembler. You might need to slightly modify the code to adapt it to your compiler/assembler specifics (#pragmas, inline assembly language syntax, options, etc.). If the kernel is provided in object form only then you'll need to make sure it's compiled for the tools you intend to use. If you are developing in C or C++, make sure the compiler is ANSI compliant and that the code generates reentrant code. In fact, reentrancy is very important if you use a real-time kernel.

Modern tools now come with Integrated Development Environments (IDEs) allowing you to create, compile/assemble and test the code all for a single interface. A number of debuggers come with kernel awareness built-in. A kernel aware debugger knows the contents of the kernel data structures. As you step through your code, you can see the task status of all your task update in special windows. These windows generally display which tasks are running, which tasks are waiting for events and which event they are waiting for, how much stack space is being used by each task and so on. You might want to check with your tool vendor to see whether they provide kernel awareness and if so, which kernels they support.

Components If your application requires communication protocols such as TCP/IP, services, libraries or other components such as a Graphical User Interface (GUI), USB, CAN, a file system and more, consider their availability.

Conclusion

A kernel is a very useful component to add to your embedded system. A kernel allows you to prioritize the work that needs to be done in separate tasks. Kernels also provide you with services that allow you to better manage the interaction between tasks. The responsiveness to high priority tasks is virtually unaffected when you add low priority tasks.

You should consider purchasing a commercial kernel instead of 'rolling your own'. Even though the concepts are simple and straightforward, kernels take many man-months and even man-years to develop and fully debug. An in-house kernel is also never done. There is also never enough time to properly document the in-house kernel and thus, you might find yourself at the mercy of one or two employees. Employees could leave your company and then you could be stuck with a product that nobody can or wants to support.

Reentrancy

Jack Ganssle

Virtually every embedded system uses interrupts; many support multitasking or multithreaded operations. These sorts of applications can expect the program's control flow to change contexts at just about any time. When that interrupt comes, the current operation is put on hold and another function or task starts running. What happens if functions and tasks share variables? Disaster surely looms if one routine corrupts the other's data.

By carefully controlling how data is shared, we create "reentrant" functions, those that allow multiple concurrent invocations that do not interfere with each other. The word "pure" is sometimes used interchangeably with "reentrant."

Reentrancy was originally invented for mainframes, in the days when memory was a valuable commodity. System operators noticed that a dozen or hundreds of identical copies of a few big programs would be in the computer's memory array at any time. At the University of Maryland, my old hacking grounds, the monster Univac 1108 had one of the early reentrant FORTRAN compilers. It burned up a (for those days) breathtaking 32kw of system memory, but being reentrant, it required only 32k even if 50 users were running it. Everyone executed the same code, from the same set of addresses. Each person had his or her own data area, yet everyone running the compiler quite literally executed identical code. As the operating system changed contexts from user to user it swapped data areas so one person's work didn't affect any other. Share the code, but not the data.

In the embedded world a routine must satisfy the following conditions to be reentrant:

1. It uses all shared variables in an atomic way, unless each is allocated to a specific instance of the function.

2. It does not call non-reentrant functions.

3. It does not use the hardware in a non-atomic way.

Atomic Variables

Both the first and last rules use the word "atomic," which comes from the Greek word meaning "indivisible." In the computer world "atomic" means an operation that cannot be interrupted. Consider the assembly language instruction:

```
mov ax,bx
```

Since nothing short of a reset can stop or interrupt this instruction it's atomic. It will start and complete without any interference from other tasks or interrupts.

The first part of rule 1 requires the atomic use of shared variables. Suppose two functions each share the global variable "foobar." Function A contains:

```
temp=foobar;
temp+=1;
foobar=temp;
```

This code is not reentrant, because foobar is used non-atomically. That is, it takes three statements to change its value, not one. The foobar handling is not indivisible; an interrupt can come between these statements, switch context to the other function, which then may also try and change foobar. Clearly there's a conflict; foobar will wind up with an incorrect value, the autopilot will crash and hundreds of screaming people will wonder "why didn't they teach those developers about reentrancy?"

Suppose, instead, function A looks like:

```
foobar+=1;
```

Now the operation is atomic; an interrupt will not suspend processing with foobar in a partially-changed state, so the routine is reentrant.

Except... do you really know what your C compiler generates? On an x86 processor the code might look like:

```
mov ax, [foobar]
inc ax
mov [foobar],ax
```

which is clearly not atomic, and so not reentrant. The atomic version is:

```
inc [foobar]
```

The moral is to be wary of the compiler; assume it generates atomic code and you may find 60 Minutes knocking at your door.

The second part of the first reentrancy rule reads "...unless each is allocated to a specific instance of the function." This is an exception to the atomic rule that skirts the issue of shared variables.

An "instance" is a path through the code. There's no reason a single function can't be called from many other places. In a multitasking environment it's quite possible that several copies of the function may indeed be executing concurrently. (Suppose the routine is a driver that retrieves data from a queue; many different parts of the code may want queued data more or less simultaneously). Each execution path is an "instance" of the code.

Consider:

```
int foo;
void some_function(void){
foo++;
}
```

foo is a global variable whose scope exists beyond that of the function. Even if no other routine uses foo, some_function can trash the variable if more than one instance of it runs at any time.

C and C++ can save us from this peril. Use automatic variables. That is, declare foo inside of the function. Then, each instance of the routine will use a new version of foo created from the stack, as follows:

```
void some_function(void){
int foo;
foo++;
}
```

Another option is to dynamically assign memory (using malloc), again so each incarnation uses a unique data area. The fundamental reentrancy problem is thus avoided, as it's impossible for multiple instances to stamp on a common version of the variable.

Two More Rules

The rest of the rules are very simple.

Rule 2 tells us a calling function inherits the reentrancy problems of the callee. That makes sense; if other code inside the function trashes shared variables, the system is going to crash. Using a compiled language, though, there's an insidious problem. Are you sure—really sure—that the runtime package is reentrant? Obviously string operations and a lot of other complicated things use runtime calls to do the real work. An awful lot of compilers also generate runtime calls to do, for instance, long math, or even integer multiplications and divisions.

If a function must be reentrant, talk to the compiler vendor to ensure that the entire runtime package is pure. If you buy software packages (like a protocol stack) that may be called from several places, take similar precautions to ensure the purchased routines are also reentrant.

Rule 3 is a uniquely embedded caveat. Hardware looks a lot like a variable; if it takes more than a single I/O operation to handle a device, reentrancy problems can develop.

Consider Zilog's SCC serial controller. Accessing any of the device's internal registers requires two steps: first write the register's address to a port, then read or write the register from the same port, the same I/O address. If an interrupt comes between setting the port and accessing the register another function might take over and access the device. When control returns to the first function the register address you set will be incorrect.

Keeping Code Reentrant

What are our best options for eliminating non-reentrant code? The first rule of thumb is to avoid shared variables. Globals are the source of no end of debugging woes and failed code. Use automatic variables or dynamically allocated memory.

Yet globals are also the fastest way to pass data around. It's not entirely possible to eliminate them from real time systems. So, when using a shared resource (variable or hardware) we must take a different sort of action.

The most common approach is to disable interrupts during non-reentrant code. With interrupts off the system suddenly becomes a single-process environment. There will be no context switches. Disable interrupts, do the non-reentrant work, and then turn interrupts back on.

Most times this sort of code looks like:

```
long i;
void do_something(void){
    disable_interrupts();
    i+=0x1234;
    enable_interrupts();
}
```

This solution *does not work*. If do_something() is a generic routine, perhaps called from many places, and is invoked with interrupts disabled, it return after turning them back on. The machine's context is changed, probably in a very dangerous manner.

Don't use the old excuse "yeah, but I wrote the code and I'm careful. I'll call the routine only when I know that interrupts will be on." A future programmer probably does not know about this restriction, and may see do_something() as just the ticket needed to solve some other problem… perhaps when interrupts are off.

Better code looks like:

```
long i;
void do_something(void){
    push interrupt state;
    disable_interrupts();
    i+=0x1234;
    pop interrupt state;
}
```

Shutting interrupts down does increase system latency, reducing its ability to respond to external events in a timely manner. A kinder, gentler approach is to use a semaphore to indicate when a resource is busy. Semaphores are simple on-off state indicators whose processing is inherently atomic, often used as "in-use" flags to have routines idle when a shared resource is not available.

Nearly every commercial real time operating system includes semaphores; if this is your way of achieving reentrant code, by all means use an RTOS.

Don't have an RTOS? Sometimes I see code that assigns an in-use flag to protect a shared resource, like this:

```
while (in_use);   //wait till resource free

in_use=TRUE;      //set resource busy

Do non-reentrant stuff

in_use=FALSE;     //set resource available
```

If some other routine has access to the resource it sets in_use true, cause this routine to idle till in_use gets released. Seems elegant and simple... but it does not work. An interrupt that occurs after the while statement will preempt execution. This routine feels it now has exclusive access to the resource, yet hasn't had a chance to set in_use true. Some other routine can now get access to the resource.

Some processors have a test-and-set instruction, which acts like the in-use flag, but which is interrupt-safe. It'll always work. The instruction looks something like:

```
Tset  variable ; if (variable==0){
               ;        variable=1;
               ;        returns TRUE;}
               ;  else {returns FALSE;}
```

If you're not so lucky to have a test-and-set, try the following:

```
loop:     mov   al,0           ; 0 means "in use"
          xchg  al,variable
          cmp   al,0
          je    loop           ; loop if in use
```

If al=0, we swapped 0 with zero; nothing changed, but the code loops since someone else is using the resource. If al=1, we put a 0 into the "in use" variable, marking the resource as busy. We fall out of the loop, now having control of the resource. It'll work every time.

Recursion

No discussion of reentrancy is complete without mentioning recursion, if only because there's so much confusion between the two.

A function is recursive if it calls itself. That's a classic way to remove iteration from many sorts of algorithms. Given enough stack space this is a perfectly valid—though tough to debug—way to write code. Since a recursive function calls itself, clearly it must be reentrant to avoid trashing its variables. So all recursive functions must be reentrant... but not all reentrant functions are recursive.

Asynchronous Hardware/Firmware

But there are subtler issues that result from the interaction of hardware and software. These may not meet the classical definition of reentrancy, but pose similar risks, and require similar solutions.

We work at that fuzzy interface between hardware and software, which creates additional problems due to the interactions of our code and the device. Some cause erratic and quite impossible-to-diagnose crashes that infuriate our customers. The worst bugs of all are those that appear infrequently, that can't be reproduced. Yet a reliable system just cannot tolerate any sort of defect, especially the random one that passes our tests, perhaps dismissed with the "ah, it's just a glitch" behavior.

Potential evil lurks whenever hardware and software interact asynchronously. That is, when some physical device runs at its own rate, sampled by firmware running at some different speed.

I was poking through some open-source code and came across a typical example of asynchronous interactions. The RTEMS real time operating system provided by OAR Corporation (ftp://ftp.oarcorp.com/pub/rtems/releases/4.5.0/ for the code) is a nicely written, well organized product with a lot of neat features. But the timer handling routines, at least for the 68302 distribution, are flawed in a way that will fail infrequently but possibly catastrophically. This is just one very public example of the problem I constantly see in buried in proprietary firmware.

The code is simple and straightforward, and looks much like any other timer handler.

```
int timer_hi;
interrupt timer(){
    ++timer_hi;}

long read_timer(void){
    unsigned int low, high;
    low =inword(hardware_register);
    high=timer_hi;
    return (high<<16 + low);}
```

There's an interrupt service routine invoked when the 16 bit hardware timer overflows. The ISR services the hardware, increments a global variable named timer_hi, and returns. So timer_hi maintains the number of times the hardware counted to 65536.

Function read_timer returns the current "time" (the elapsed time in microseconds as tracked by the ISR and the hardware timer). It, too, is delightfully free of complications. Like most of these sorts of routines it reads the current contents of the hardware's timer register, shifts Timer_hi left 16 bits, and adds in the value read from the timer. That is, the current time is the concatenation the timer's current value and the number of overflows.

Suppose the hardware rolled over 5 times, creating five interrupts. `timer_hi` equals 5. Perhaps the internal register is, when we call `read_timer`, 0x1000. The routine returns a value of 0x51000. Simple enough and seemingly devoid of problems.

Race Conditions

But let's think about this more carefully. There's really two things going on at the same time. *Not* concurrently, which means "apparently at the same time," as in a multitasking environment where the RTOS doles out CPU resources so all tasks *appear* to be running simultaneously. No, in this case the code in `read_timer` executes whenever called, and the clock-counting timer runs at its own rate. The two are asynchronous.

A fundamental rule of hardware design is to panic whenever asynchronous events suddenly synchronize. For instance, when two different processors share a memory array there's quite a bit of convoluted logic required to ensure that only one gets access at any time. If the CPUs use different clocks the problem is much trickier, since the designer may find the two requesting exclusive memory access within fractions of a nanosecond of each other. This is called a "race" condition and is the source of many gray hairs and dramatic failures.

One of `read_timer`'s race conditions might be:

- It reads the hardware and gets, let's say, a value of 0xffff.

- Before having a chance to retrieve the high part of the time from variable `timer_hi`, the hardware increments again to 0x0000.

- The overflow triggers an interrupt. The ISR runs. `timer_hi` is now 0x0001, not 0 as it was just nanoseconds before.

- The ISR returns, our fearless `read_timer` routine, with no idea an interrupt occurred, blithely concatenates the new 0x0001 with the previously-read timer value of 0xffff, and returns 0x1ffff—a hugely incorrect value.

Or, suppose `read_timer` is called during a time when interrupts are disabled—say, if some other ISR needs the time. One of the few perils of writing encapsulated code and drivers is that you're never quite sure what state the system is in when the routine gets called. In this case:

- `read_timer` starts. The timer is 0xffff with no overflows.

- Before much else happens it counts to 0x0000. With interrupts off the pending interrupt gets deferred.

- `read_timer` returns a value of 0x0000 instead of the correct 0x10000, or the reasonable 0xffff.

So the algorithm that seemed so simple has quite subtle problems, necessitating a more sophisticated approach. The RTEMS RTOS, at least in its 68k distribution, will likely create infrequent but serious errors.

Sure, the odds of getting a mis-read are small. In fact, the chance of getting an error plummets as the frequency we call `read_timer` decreases. How often will the race condition surface? Once a week? Monthly?

Many embedded systems run for years without rebooting. Reliable products must *never* contain fragile code. Our challenge as designers of robust systems is to identify these sorts of issues and create alternative solutions that work correctly, every time.

Options

Fortunately a number of solutions do exist. The easiest is to stop the timer before attempting to read it. There will be no chance of an overflow putting the upper and lower halves of the data out of sync. This is a simple and guaranteed solution.

We will lose time. Since the hardware generally counts the processor's clock, or clock divided by a small number, it may lose quite a few ticks during the handful of instructions executed to do the reads. The problem will be much worse if an interrupt causes a context switch after disabling the counting. Turning interrupts off during this period will eliminate unwanted tasking, but increases both system latency and complexity.

I just *hate* disabling interrupts; system latency goes up and sometimes the debugging tools get a bit funky. When reading code a red flag goes up if I see a lot of disable interrupt instructions sprinkled about. Though not necessarily bad, it's often a sign that either the code was beaten into submission (made to work by heroic debugging instead of careful design), or there's something quite difficult and odd about the environment.

Another solution is to read the `timer_hi` variable, then the hardware timer, and then re-read `timer_hi`. An interrupt occurred if both variable values aren't identical. Iterate till the two variable reads are equal. The upside: correct data, interrupts stay on, and the system doesn't lose counts.

The downside: in a heavily-loaded, multitasking environment, it's possible that the routine could loop for rather a long time before getting two identical reads. The function's execution time is non-deterministic. We've gone from a very simple timer reader to somewhat more complex code that could run for milliseconds instead of microseconds.

Another alternative might be to simply disable interrupts around the reads. This will prevent the ISR from gaining control and changing `timer_hi` after we've already read it, but creates another issue.

We enter `read_timer` and immediately shut down interrupts. Suppose the hardware timer is at our notoriously-problematic 0xffff, and `timer_hi` is zero. Now, before the code has a chance to do anything else, the overflow occurs. With context switching shut down we miss the rollover. The code reads a zero from both the timer register and from `timer_hi`, returning zero instead of the correct 0x10000, or even a reasonable 0x0ffff.

Yet disabling interrupts is probably indeed a good thing to do, despite my rant against this practice. With them on there's always the chance our reading routine will be suspended by higher priority tasks and other ISRs for perhaps a very long time. Maybe long enough for the timer to roll over several times. So let's try to fix the code. Consider the following:

```
long Read_timer(void){
    unsigned int low, high;
    push_interrupt_state;
    disable_interrupts;
    low=inword(Timer_register);
    high=timer_hi;
    if(inword(timer_overflow))
       {++high;
        low=inword(timer_register);}
    pop_interrupt_state;
    return (((ulong)high)<<16 + (ulong)low);
}
```

We've made three changes to the RTEMS code. First, interrupts are off, as described.

Second, you'll note that there's no explicit interrupt re-enable. Two new pseudo-C statements have appeared which push and pop the interrupt state. Trust me for a moment—this is just a more sophisticated way to manage the state of system interrupts.

The third change is a new test that looks at something called "`timer_overflow`," an input port that is part of the hardware. Most timers have a testable bit that signals an overflow took place. We check this to see if an overflow occurred between turning interrupts off and reading the low part of the time from the device. With an inactive ISR variable `timer_hi` won't properly reflect such an overflow.

We test the status bit and reread the hardware count if an overflow had happened. Manually incrementing the high part corrects for the suspended ISR. The code then concatenates the two fixed values and returns the correct result. Every time.

With interrupts off we have increased latency. However, there are no loops; the code's execution time is entirely deterministic.

Other RTOSes

Unhappily, race conditions occur anytime we're need more than one read to access data that's changing asynchronously to the software. If you're reading X and Y coordinates, even with just 8 bits of resolution, from a moving machine there's some peril they could be seriously out of sync if two reads are required. A ten bit encoder managed through byte-wide ports potentially could create a similar risk.

Having dealt with this problem in a number of embedded systems over the years, I wasn't too shocked to see it in the RTEMS RTOS. It's a pretty obscure issue, after all, though terribly real and potentially deadly. For fun I looked through the source of uC/OS, another very popular operating system whose source is on the net (see www.ucos-ii.com). uC/OS never reads the timer's hardware. It only counts overflows as detected by the ISR, as there's no need for higher resolution. There's no chance of an incorrect value.

Some of you, particularly those with hardware backgrounds, may be clucking over an obvious solution I've yet to mention. Add an input capture register between the timer and the system; the code sets a "lock the value into the latch" bit, then reads this safely unchanging data. The register is nothing more than a parallel latch, as wide as the input data. A single clock line drives each flip-flop in the latch; when strobed it locks the data into the register. The output is fed to a pair of processor input ports.

When it's time to read a safe, unchanging value the code issues a "hold the data now" command which strobes encoder values into the latch. So all bits are stored and can be read by the software at any time, with no fear of things changing between reads.

Some designers tie the register's clock input to one of the port control lines. The I/O read instruction then automatically strobes data into the latch, assuming one is wise enough to ensure the register latches data on the leading edge of the clock.

The input capture register is a very simple way to suspend moving data during the duration of a couple of reads. At first glance it seems perfectly safe. But a bit of analysis shows that for asynchronous inputs it *is not reliable*. We're using hardware to fix a software problem, so we must be aware of the limitations of physical logic devices.

To simplify things for a minute, let's zoom in on that input capture register and examine just one of its bits. Each gets stored in a flip-flop, a bit of logic that might have only three connections: data in, data out, and clock. When the input is a one, strobing clock puts a one at the output.

But suppose the input changes at about the same time clock cycles? What happens? The short answer is that no one knows.

Metastable States

Every flip-flop has two critical specifications we violate at our peril. "Set-up time" is the minimum number of nanoseconds that input data must be stable *before* clock comes. "Hold time" tells us how long to keep the data present *after* clock transitions. These specs vary depending on the logic device. Some might require tens of nanoseconds of set-up and/or hold time; others need an order of magnitude less.

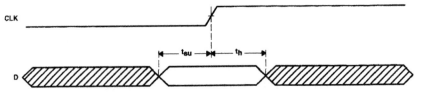

Figure 16-1: Setup and hold times.

If we tend to our knitting we'll respect these parameters and the flip-flop will always be totally predictable. But when things are asynchronous—say, the wrist rotates at it's own rate and the software does a read whenever it needs data—there's a chance the we'll violate set-up or hold time.

Suppose the flip-flop requires 3 nanoseconds of set-up time. Our data changes within that window, flipping state perhaps a single nanosecond before clock transitions. The device will go into a metastable state where the output gets very strange indeed.

By violating the spec the device really doesn't know if we presented a zero or a one. It's output goes, not to a logic state, but to either a half-level (in between the digital norms) or it will oscillate, toggling wildly between states. The flip-flop is metastable.

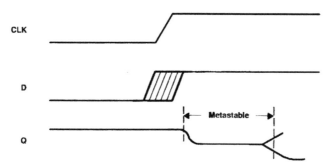

Figure 16-2: A metastable state.

This craziness doesn't last long; typically after a few to 50 nanoseconds the oscillations damp out or the half-state disappears, leaving the output at a valid one or zero. But which one is it? This is a digital system, and we expect ones to be ones, and zeroes zeroes.

The output is *random*. Bummer, that. You cannot predict which level it will assume. That sure makes it hard to design predictable digital systems!

Hardware folks feel that the random output isn't a problem. Since the input changed at almost exactly the same time the clock strobed, either a zero or a one is reasonable. If we had clocked just a hair ahead or behind we'd have gotten a different value, anyway. Philosophically, who knows which state we measured? Is this really a big deal? Maybe not to the EEs, but this impacts our software in a big way, as we'll see shortly.

Metastability occurs only when clock and data arrive almost simultaneously; the odds increase as clock rates soar. An equally important factor is the type of logic component used; slower logic (like 74HCxx) has a much wider metastable window than faster devices (say, 74FCTxx). Clearly at reasonable rates the odds of the two asynchronous signals arriving closely enough in time to cause a metastable situation are low; measurable, yes, important, certainly. With a 10 MHz clock and 10 KHz data rate, using typical but not terribly speedy logic, metastable errors occur about once a minute. Though infrequent, no reliable system can stand that failure rate.

The classic metastable fix uses two flip flops connected in series. Data goes to the first; its output feeds the data input of the second. Both use the same clock input. The second flop's output will be "correct" after two clocks, since the odds of two metastable events occurring back-to-back are almost nil. With two flip-flops, at reasonable data rates errors occur millions or even billions of years apart. Good enough for most systems.

But "correct" means the second stage's output will not be metastable: it's not oscillating, nor is it at an illegal voltage level. There's still an equal chance the value will be in either legal logic state.

Firmware, not Hardware

To my knowledge there's no literature about how metastability affects software, yet it poses very real threats to building a reliable system.

Hardware designers smugly cure their metastability problem using the two stage flops described. Their domain is that of a single bit, whose input changed just about the same time the clock transition. Thinking in such narrow terms it's indeed reasonable to accept the inherent random output the flops generate.

But we software folks are reading parallel I/O ports, each perhaps 8 bits wide. That means there are 8 flip-flops in the input capture register, all driven by the same clock pulse.

Let's look at what might happen. The encoder changes from 0xff to 0x100. This small difference might represent just a tiny change in angle. We request a read at just about the same time the data changes; our input operation strobes the capture register's clock creating a violation of set-up or hold time. Every input bit changes; each of the flip flops inside the register goes metastable. After a short time the oscillations die out, but now every bit in the register is random. Though the hardware folks might shrug and complain that no one knows what the right value was, since everything changed as clock arrived, in fact the data was around 0xff or 0x100. A random result of, say, 0x12 is absurd and totally unacceptable, and may lead to crazy system behavior.

The case where data goes from 0xff to 0x100 is pathological since every bit changes at once. The system faces the same peril whenever lots of bits change. 0x0f to 0x10. 0x1f to 0x20. The upper, unchanging data bits will always latch correctly; but every changing bit is at risk.

Why not use the multiple flip-flop solution? Connect two input capture registers in series, both driven by the same clock. Though this will eliminate the illegal logic states and oscillations, the second stage's output will be random as well.

One option is to ignore metastability and hope for the best. Or use very fast logic with very narrow set-up/hold time windows to reduce the odds of failure. If the code samples in the inputs infrequently it's possible to reduce metastability to one chance in millions or even billions. Building a safety critical system? Feeling lucky?

It is possible to build a synchronizer circuit that takes a request for a read from the processor, combines it with a data available bit from the I/O device, responding with a data-OK signal back to the CPU. This is non-trivial and prone to errors.

An alternative is to use a different coding scheme for the I/O device. Buy an encoder with Gray code output, for example (if you can find one). Gray code is a counting scheme where only a single bit changes between numbers, as follows:

0	000
1	001
2	011
3	010
4	110
5	111
6	101
7	100

Gray code makes sense if, and only if, your code reads the device faster than it's likely to change, and if the changes happen in a fairly predictable fashion—like counting up. Then there's no real chance of more than a single bit changing between reads; if the inputs go metastable only one bit will be wrong. The result will still be reasonable.

Another solution is to compute a parity or checksum of the input data before the capture register. Latch that, as well, into the register. Have the code compute parity and compare it to that read; if there's an error do another read.

Though I've discussed adding an input capture register, please don't think that this is the root cause of the problem. Without that register—if you just feed the asynchronous inputs directly into the CPU—it's quite possible to violate the processor's innate set-up/hold times. There's no free lunch; all logic has physical constraints we must honor.

Some designs will never have a metastability problem. It always stems from violating set-up or hold times, which in turn comes from either poor design or asynchronous inputs.

All of this discussion has revolved around asynchronous inputs, when the clock and data are unrelated in time. Be wary of anything not slaved to the processor's clock. Interrupts are a notorious source of problems. If caused by, say, someone pressing a button, be sure that the interrupt itself, and the vector-generating logic, don't violate the processor's set-up and hold times.

But in computer systems most things do happen synchronously. If you're reading a timer that operates from the CPU's clock, it is inherently synchronous to the code. From a metastability standpoint it's totally safe.

Bad design, though, can plague any electronic system. Every logic component takes time to propagate data; when a signal traverses many devices the delays can add up significantly. If the data then goes to a latch it's quite possible that the delays may cause the input to transition at the same time as the clock. Instant metastability.

Designers are pretty careful to avoid these situations, though. Do be wary of FPGAs and other components where the delays vary depending on how the software routes the device. And when latching data or clocking a counter it's not hard to create a metastability problem by using the wrong clock edge. Pick the edge that gives the device time to settle before it's read.

What about analog inputs? Connect a 12 bit A/D converter to two 8 bit ports and we'd seem to have a similar problem: the analog data can wiggle all over, changing during the time we read the two ports. However, there's no need for an input capture register because the converter itself generally includes a "sample and hold" block, which stores the analog signal while the A/D digitizes. Most A/Ds then store the digital value till we start the next conversion.

Other sorts of inputs we use all share this problem. Suppose a robot uses a 10 bit encoder to monitor the angular location of a wrist joint. As the wrist rotates the encoder sends back a binary code, 10 bits wide, representing the joint's current position. An 8 bit processor requires two distinct I/O instructions—two byte-wide reads—to get the data. No matter how fast the computer might be there's a finite time between the reads during which the encoder data may change.

The wrist is rotating. A "get_position" routine reads 0xff from the low part of the position data. Then, before the next instruction, the encoder rolls over to 0x100. "get_position" reads the high part of the data—now 0x1—and returns a position of 0x1ff, clearly in error and perhaps even impossible.

This is a common problem. Handling input from a two axis controller? If the hardware continues to move during our reads, then the X and Y data will be slightly uncorrelated, perhaps yielding impossible results. One friend tracked a rare autopilot failure to the way the code read a flux-gate compass, whose output is a pair of related quadrature signals. Reading them at disparate times, while the vessel continued to move, yielded impossible heading data.

Interrupt Latency

Jack Ganssle

My dad was a mechanical engineer, who spent his career designing spacecraft. I remember back even in the early days of the space program how he and his colleagues analyzed seemingly every aspect of their creations' behavior. Center of gravity calculations insured that the vehicles were always balanced. Thermal studies guaranteed nothing got too hot or too cold. Detailed structural mode analysis even identified how the system would vibrate, to avoid destructive resonances induced by the brutal launch phase.

Though they were creating products that worked in a harsh and often unknown environment, their detailed computations profiled how the systems would behave.

Think about civil engineers. Today no one builds a bridge without "doing the math." That delicate web of cables supporting a thin dancing roadway is simply going to work. Period. The calculations proved it long before contractors started pouring concrete.

Airplane designers also use quantitative methods to predict performance. When was the last time you heard of a new plane design that wouldn't fly? Yet wing shapes are complex and notoriously resistant to analytical methods. In the absence of adequate theory, the engineers rely on extensive e tables acquired over decades of wind tunnel experiments. The engineers can still understand how their product will work—in general—before bending metal.

Compare this to our field. Despite decades of research, formal methods to prove software correctness are still impractical for real systems. We embedded engineers build, then test, with no real proof that our products will work. When we pick a CPU, clock speed, memory size, we're betting that our off-the-cuff guesses will be adequate when, a year later, we're starting to test 100,000+ lines of code. Experience plays an important role in getting the resource requirements right. All too often luck is even more critical. But hope is our chief tool, and the knowledge that generally with enough heroics we can overcome most challenges.

In my position as embedded gadfly, looking into thousands of projects, I figure some 10–15% are total failures due simply to the use of inadequate resources. The 8051 just can't handle that firehose of data. The PowerPC part was a good choice but the program grew to twice the size of available Flash… and with the new cost model the product is not viable.

Recently I've been seeing quite a bit written about ways to make our embedded systems more predictable, to insure they react fast enough to external stimuli, to guarantee processes complete on-time. To my knowledge there is no realistically useful way to calculate predictability. In most cases we build the system and start changing stuff if it runs too slowly. Compared to aerospace and civil engineers we're working in the dark.

It's especially hard to predict behavior when asynchronous activities alter program flow. Multitasking and interrupts both lead to impossible-to-analyze problems. Recent threads on USENET, as well as some discussions at the Embedded Systems Conference, suggest banning interrupts altogether! I guess this does lead to a system that's easier to analyze, but the solution strikes me as far too radical.

I've built polled systems. Yech. Worse are applications that must deal with several different things, more or less concurrently, without using multitasking. The software in both situations is invariably a convoluted mess. About 20 years ago I naively built a steel thickness gauge without an RTOS, only to later have to shoehorn one in. There were too many async things going on; the in-line code grew to outlandish complexity. I'm still trying to figure out how to explain that particular sin to St. Peter.

A particularly vexing problem is to ensure the system will respond to external inputs in a timely manner. How can we guarantee that an interrupt will be recognized and processed fast enough to keep the system reliable?

Let's look in some detail at the first of the requirements: that an interrupt be recognized in-time. Simple enough, it seems. Page through the processor's databook and you'll find a spec called "latency," a number always listed at sub-microsecond levels. No doubt a footnote defines latency as the longest time between when the interrupt occurs and when the CPU suspends the current processing context. That would seem to be the interrupt response time... but it ain't.

Latency as defined by CPU vendors varies from zero (the processor is ready to handle an interrupt RIGHT NOW) to the max time specified. It's a product of what sort of instruction is going on. Obviously it's a bad idea to change contexts in the middle of executing an instruction, so the processor generally waits till the current instruction is complete before sampling the interrupt input. Now, if it's doing a simple register to register move that may be only a single clock cycle, a mere 50 nsec on a zero wait state 20-MHz processor. Not much of a delay at all.

Other instructions are much slower. Multiplies can take dozens of clocks. Read-modify-write instructions (like "increment memory") are also inherently pokey. Max latency numbers come from these slowest of instructions.

Many CPUs include looping constructs that can take hundreds, even thousands of microseconds. A block memory-to-memory transfer, for instance, initiated by a single instruction, might run for an awfully long time, driving latency figures out of sight. All processors I'm aware of will accept an interrupt in the middle of these long loops to keep interrupt response reasonable. The block move will be suspended, but enough context is saved to allow the transfer to resume when the ISR (Interrupt Service Routine) completes.

So, the latency figure in the datasheet tells us the longest time the processor can't service interrupts. The number is totally useless to firmware engineers.

OK, if you're building an extreme cycle-countin', nanosecond-poor, gray-hair-inducing system then perhaps that 300 nsec latency figure is indeed a critical part of your system's performance. For the rest of us, real latency—the 99% component of interrupt response—comes not from what the CPU is doing, but from our own software design. And that, my friend, is hard to predict at design time. Without formal methods we need empirical ways to manage latency.

If latency is time between getting an interrupt and entering the ISR, then surely most occurs because we've disabled interrupts! It's because of the way we wrote the darn code. Turn interrupts off for even a few C statements and latency might run to hundreds of microseconds, far more than those handful of nanoseconds quoted by CPU vendors.

No matter how carefully you build the application, you'll be turning interrupts off frequently. Even code that never issues a "disable interrupt" instruction does, indeed, disable them often. For, every time a hardware event issues an interrupt request, the processor itself does an automatic disable, one that stays in effect till you explicitly re-enable them inside of the ISR. Count on skyrocketing latency as a result.

Of course, on many processors we don't so much as turn interrupts off as change priority levels. A 68K receiving an interrupt on level 5 will prohibit all interrupts at this and lower levels till our code explicitly re-enables them in the ISR. Higher priority devices will still function, but latency for all level 1 to 5 devices is infinity till the code does its thing.

So, in an ISR re-enable interrupts as soon as possible. When reading code one of my "rules of thumb" is that code which does the enable just before the return is probably flawed. Most of us were taught to defer the interrupt enable till the end of the ISR. But that prolongs latency unacceptably. Every other interrupt (at least at or below that priority level) will be shut down till the ISR completes. Better: enter the routine, do all of the non-reentrant things (like handling hardware), and *then* enable interrupts. Run the rest of the ISR, which manages reentrant variables and the like, with interrupts on. You'll reduce latency and increase system performance.

The downside might be a need for more stack space if that same interrupt can re-invoke itself. There's nothing wrong with this in a properly designed and reentrant ISR, but the stack will grow till all pending interrupts get serviced.

The second biggest cause of latency is excessive use of the disable interrupts instruction. Shared resources—global variables, hardware, and the like—will cause erratic crashes when two asynchronous activities try to access them simultaneously. It's up to us to keep the code reentrant by either keeping all such accesses atomic, or by limiting access to a single task at a time. The classic approach is to disable interrupts around such accesses. Though a simple solution, it comes at the cost of increased latency.

Taking Data

So what is the latency of your system? Do you know? Why not?

It's appalling that so many of us build systems with a "if the stupid thing works at all, ship it" philosophy. It seems to me there are certain critical parameters we must understand in order to properly develop and maintain a product. Like, is there any free ROM space? Is the system 20% loaded... or 99%? How bad is the max latency?

Latency is pretty easy to measure; sometimes those measurements will yield surprising and scary results.

Perhaps the easiest way to get a feel for interrupt response is to instrument each ISR with an instruction that toggles a parallel output bit *high* when the routine starts. Drive it low just as it exits. Connect this bit to one input of an oscilloscope, tying the other input to the interrupt signal itself.

The amount of information this simple setup gives is breathtaking. Measure time from the assertion of the interrupt till the parallel bit goes high. That's latency, minus a bit for the overhead of managing the instrumentation bit. Twiddle the scope's time base to measure this to any level of precision required.

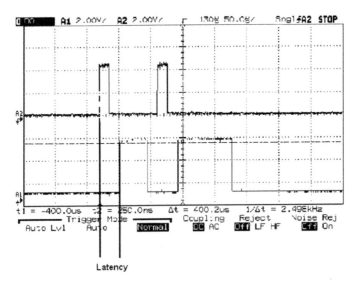

Figure 17-1: The latency is the time from when the interrupt signal appears, till the ISR starts.

The time the bit stays high is the ISR's total execution time. Tired of guessing how fast your code runs? This is quantitative, cheap, and accurate.

In a real system, interrupts come often. Latency varies depending on what other things are going on. Use a digital scope in storage mode. After the assertion of the interrupt input you'll see a clear space—that's the minimum system latency to this input. Then there will be hash, a blur as the instrumentation bit goes high at different times relative to the interrupt input. These represent variations in latency. When the blur resolves itself into a solid *high*, that's the maximum latency.

All this, for the mere cost of one unused parallel bit.

If you've got a spare timer channel, there's another approach that requires neither extra bits nor a scope. Build an ISR just for measurement purposes that services interrupts from the timer.

On initialization, start the timer counting up, programmed to interrupt when the count overflows. Have it count as fast as possible. Keep the ISR dead simple, with minimal overhead. This is a good thing to write in assembly language to minimize unneeded code. Too many C compilers push *everything* inside interrupt handlers.

The ISR itself reads the timer's count register and sums the number into a long variable, perhaps called `total_time`. Also increment a counter (`iterations`). Clean up and return.

The trick here is that, although the timer reads zero when it tosses out the overflow interrupt, the timer register continues counting even as the CPU is busy getting ready to invoke the ISR. If the system is busy processing another interrupt, or perhaps stuck in an interrupt-disabled state, the counter continues to increment. An infinitely fast CPU with no latency would start the instrumentation ISR with the counter register equal to zero. Real processors with more usual latency issues will find the counter at some positive non-zero value that indicates how long the system was off doing other things.

So, average latency is just the time accumulated into `total_time` (normalized to microseconds) divided by the number of times the ISR ran (`iterations`).

It's easy to extend the idea to give even more information. Possibly the most important thing was can know about our interrupts is the longest latency. Add a few lines of code to compare for and log the max time.

Is the method perfect? Of course not. The data is somewhat statistical, so can miss single-point outlying events. Very speedy processors may run so much faster than the timer tick rate that they always log latencies of zero… though this may indicate that for all practical purposes latencies are short enough to not be significant.

The point is that knowledge is power; once we understand the magnitude of latency reasons for missed interrupts become glaringly apparent.

Try running these experiments on purchased software components. One embedded DOS, running on a 100-MHz 486, yielded latencies in the tens of milliseconds!

Understanding Your C Compiler:
How to Minimize Code Size

Jakob Engblom

Jakob Engblom got his PhD in computer systems from Uppsala University in 2002, and his MSc in Computer Science from Uppsala University in 1997. Jakob is currently a senior developer at Virtutech, and an adjunct professor at the department of Information Technology at Uppsala University. His interests include embedded systems programming, computer architecture, simulation technology, real-time systems, and compiler technology. Prior to joining Virtutech, Jakob worked for IAR Systems as a speaker, researcher, and compiler developer. For more information, see http://user.it.uu.se/~jakob.

A C compiler is a basic tool for most embedded systems programmers. It is the tool by which the ideas and algorithms in your application (expressed as C source code) are transformed into machine code executable by your target processor. To a large extent, the C compiler determines how large the executable code for the application will be.

A compiler performs many transformations on a program in order to generate the best possible code. Examples of such transformations are storing values in registers instead of memory, removing code which does nothing useful, reordering computations in a more efficient order, and replacing arithmetic operations by cheaper operations. The C language provides the compiler with significant freedom regarding how to precisely implement each C operation on the target system. This freedom is an important factor in why C can usually be compiled very efficiently, but a programmer needs to be aware of the compiler's freedom in order to write robust code.

To most programmers of embedded systems, the case that a program does not *quite* fit into the available memory is a familiar phenomenon. Recoding parts of an application in assembly language or throwing out functionality may seem to be the only alternatives, while the solution could be as simple as rewriting the C code in a more compiler-friendly manner.

In order to write code that is compiler friendly, you need to have a working understanding of compilers. Some simple changes to a program, like changing the data type of a frequently-accessed variable, can have a big impact on code size while other changes have no effect at all. Having an idea of what a compiler can and cannot do makes such optimization work much easier.

Modern C Compilers

Assembly programs specify both what, how, and the precise order in which calculations should be carried out. A C program, on the other hand, only specifies the calculations that should be performed. With some restrictions, the order and the technique used to realize the calculations are up to the compiler.

The compiler will look at the code and try to understand what is being calculated. It will then generate the best possible code, given the information it has managed to obtain, locally within a single statement and also across entire functions and sometimes even whole programs.

The Structure of a Compiler

In general, a program is processed in six main steps in a modern compiler (not all compilers follow this blueprint completely, but as a conceptual guide it is sufficient):

- **Parser:** The conversion from C source code to an intermediate language.

- **High-level optimization:** Optimizations on the intermediate code.

- **Code generation:** Generation of target machine code from the intermediate code.

- **Low-level optimization:** Optimizations on the machine code.

- **Assembly:** Generation of an object file that can be linked from the target machine code.

- **Linking:** Linking of all the code for a program into an executable or downloadable file.

The parser parses the C source code, checking the syntax and generating error messages if syntactical errors are found in the source. If no errors are found, the parser then generates intermediate code (an internal representation of the parsed code), and compilation proceeds with the first optimization pass.

The high-level optimizer transforms the code to make it better. The optimizer has a large number of transformations available that can improve the code, and will perform those that it deems relevant to the program at hand. Note that we use the word "transformation" and not "optimization." "Optimization" is a bit of a misnomer. It conveys the intuition that a change always improves a program and that we actually find optimal solutions, while in fact optimal solutions are very expensive or even impossible to find (undecidable, in computer science lingo). To ensure reasonable compilation times and the termination of the compilation process, the compiler has to use heuristic methods ("good guesses"). Transforming a program is a highly non-linear activity, where different orderings of transformations will yield different results, and some transformations may actually make the code worse. Piling on more "optimizations" will not necessarily yield better code.

When the high-level optimizer is done, the code generator transforms the intermediate code to the target processor instruction set. This stage is performed, piece-by-piece, on the intermediate code from the optimizer, and the compiler will try to do smart things on the level of a single expression or statement, but not across several statements.

The code generator will also have to account for any differences between the C language and the target processor. For example, 32-bit arithmetic will have to be broken down to 8-bit arithmetic for a small embedded target (like an Intel 8051, Motorola 68HC08, Samsung SAM8, or Microchip PIC).

A very important part of code generation is allocating registers to variables. The goal is to keep as many values as possible in registers, since register-based operations are typically faster and smaller than memory-based operations.

After the code generator is done, another phase of optimization takes place, where transformations are performed on the target code. The low-level optimizer will clean up after the code generator (which sometimes makes suboptimal coding decisions), and perform more transformations. There are many transformations that can only be applied on the target code, and some which are repeated from the high-level phase, but on a lower level. For example, transformations like removing a "clear carry" instruction if we already know that the carry flag is zero are only possible at the target code level, since the flags are not visible before code generation.

After the low-level optimizer is finished, the code is sent to an assembler and output to an object file.

All the object files of a program are then linked to produce a final binary executable ROM image (in some format appropriate for the target). The linker may also perform some optimizations, for example by discarding unused functions.

Thus, one can see that the seemingly simple task of compiling a C program is actually a rather long and winding road through a highly complex system. Different transformations may interact, and a local improvement may be worse for the whole program. For example, an expression can typically be evaluated more efficiently if given more temporary registers. Taking a local view, it thus seems to be a good idea to provide as many registers as necessary. A global effect, however, is that variables in registers may have to be spilled to memory, which could be more expensive than evaluating the expression with fewer registers.

The Meaning of a Program

Before the compiler can apply transformations to a program, it must analyze the code to determine which transformations are legal and likely to result in improvements. The legality of transformations is determined by the semantics laid down by the C language standard.

The most basic interpretation of a C program is that only statements that have side effects or compute values used for performing side-effects are relevant to the meaning of a program. Side effects are any statements that change the global state of the program. Examples which are generally considered to be side effects are writing to a screen, changing a global variable, reading a volatile variable, and calling unknown functions.

The calculations between the side-effects are carried out according to the principle of "do what I mean, not what I say". The compiler will try to rewrite each expression into the most efficient form possible, but a rewrite is only possible if the result of the rewritten code is the same as the original expression. The C standard defines what is considered "the same," and sets the limits of allowable optimizations.

Basic Transformations

A modern compiler performs a large number of basic transformations that act locally, like folding constant expressions, replacing expensive operations by cheaper ones ("strength reduction"), removing redundant calculations, and moving invariant calculations outside of loops. The compiler can do most mechanical improvements just as well as a human programmer, but without tiring or making mistakes. The table below shows (in C form for readability) some typical basic transformations performed by a modern C compiler. Note that an important implication of this basic cleanup is that you can write code in a readable way and let the compiler calculate constant expressions and worry about using the most efficient operations.

Before Transformation	After Transformation
`unsigned short int a;` `a /= 8;`	`unsigned short int a;` `a >>= 3;   /* shift replaced divide*/`
`a *= 2;`	`a += a;   /* multiply replaced by add */` `a <<= 1;  /* or a shift */`

`a = b + c * d;`
`e = f + c * d;`
	`temp = c * d;  /* common expression */`
	`a = b + temp;  /* saves one multiply */`
	`e = f + temp;`

`a = b * 1;`	`a = b;          /* x*1 == x */`

`a = 17;`
`b = 56 + ( 2 * a );`
	`a = 17;`
	`b = 90;  /* constant value evaluated */`
	`         /* at compile time        */`

`#define BITNO 4` `port |= ( 1 << BITNO);`	`port |= 0x10; /* constant value */`

`if(a > 10)`
`{`
`   b = b * c + k;`
`   if(a < 5)`
`      a+=6;`
`}`
	`if(a > 10)`
	`{`
	`   b = b * c + k;`
	`/* unreachable code removed */`
	`}`

`a = b * c + k;` `a = k + 7;`	`/* useless computation removed */` `a = k + 7;`

`for(i=0; i<10; i++)`
`{`
`   b = k * c;`
`   p[i] = b;`
`}`
	`b = k * c;  /* constant code moved */`
	`            /* outside the loop    */`
	`for(i=0; i<10; i++)`
	`{`
	`   p[i] = b;`
	`}`

All code that is not considered useful—according to the definition in the previous section—is removed. This removal of unreachable or useless computations can cause some unexpected effects. An important example is that empty loops are completely discarded, making "empty delay loops" useless. The code shown below stopped working properly when upgrading to a modern compiler that removed useless computations:

Code that Stopped Working	... After Compiler Optimizations
| ```c
void delay(unsigned int time)
{
  unsigned int i;
  for (i=0; i<time; i++)
    ;
  return;
}

void InitHW(void)
{
  /* Highly timing-dependent code */
  OUT_SIGNAL (0x20);
  delay(120);
  OUT_SIGNAL (0x21);
  delay(121);
  OUT_SIGNAL (0x19);
}
``` | ```c
void delay(unsigned int time)
{
  /* loop removed */
  return;
}



void InitHW(void)
{
  /* Delays do not last long here… */
  OUT_SIGNAL (0x20);
  delay(120);
  OUT_SIGNAL (0x21);
  delay(121);
  OUT_SIGNAL (0x19);
}
``` |

Register Allocation

Processors usually give better performance and require smaller code when calculations are performed using registers instead of memory. Therefore, the compiler will try to assign the variables in a function to registers. A local variable or parameter will not need any RAM allocated at all if the variable can be kept in registers for the duration of the function.

If there are more variables than registers available, the compiler needs to decide which of the variables to keep in registers, and which to put in memory. This is the problem of *register allocation*, and it cannot be solved optimally. Instead, heuristic techniques are used. The algorithms used can be quite sensitive, and even small changes to a function may considerably alter the register allocation.

Note that a variable only needs a register when it is being used. If a variable is used only in a small part of a function, it will be register allocated in that part, but it will not exist in the rest of the function. This explains why a debugger sometimes tells you that a variable is "optimized away at this point."

The register allocator is limited by the language rules of C—for example, global variables have to be written back to memory when calling other functions, since they can be accessed by the called function, and all changes to global variables must be visible to all functions. Between function calls, global variables can be kept in registers.

Note that there are times when you *do not* want variables to be register allocated. For example, reading an I/O port or spinning on a lock, you want each read in the source code to be made from memory, since the variable can be changed outside the control of your program. This is where the volatile keyword is to be used. It signals to the compiler that the variable should not ever be allocated in registers, but read from memory (or written) each time it is accessed.

In general, only simple values like integers, floats, and pointers are considered for register allocation. Arrays have to reside in memory since they are designed to be accessed through pointers, and structures are usually too large. Also, on small processors, large values like 32-bit integers and floats may be hard to allocate to registers, and maybe only 16-bit and 8-bit variables will be given registers.

To help register allocation, you should strive to keep the number of simultaneously live variables low. Also, try to use the smallest possible data types for your variables, as this will reduce the number of required registers on 8-bit and 16-bit processors.

Function Calls

As assembly programmers well know, calling a function written in a high-level language can be rather complicated and costly. The calling function must save global variables back to memory, make sure to move local variables to the registers that survive the call (or save to the stack), and parameters may have to be pushed on the stack. Inside the called function, registers will have to be saved, parameters taken off the stack, and space allocated on the stack for local variables. For large functions with many parameters and variables, the effort required for a call can be quite large.

Modern compilers do their best, however, to reduce the cost of a function call, especially the use of stack space. A number of registers will be designated for parameters, so that short parameter lists will most likely be passed entirely in registers. Likewise, the return value will be put in a register, and local variables will only be put on the stack if they cannot be allocated to registers.

The number of register parameters will vary wildly between different compilers and architecture. In most cases, at least four registers are made available for parameters. Note also that just like for register allocation, only small parameter types will be passed in registers. Arrays are always passed as pointers to the array (C semantics dictate that), and structures are usually copied to the stack and the structure parameter changed to a pointer to a structure. That pointer might be passed in a register, however. To save stack space, it is thus a good idea to always use pointers to structures as parameters and not the structures themselves.

C supports functions with variable numbers of arguments. This is used in standard library functions like printf() and scanf() to provide a convenient interface. However, the implementation of variable numbers of arguments to a function incurs significant overhead. All arguments have to be put on the stack, since the function must be able to step through the parameter list using pointers to arguments, and the code accessing the arguments is much less efficient than for fixed parameter lists. There is no type-checking on the arguments, which increases the risk of bugs. Variable numbers of arguments should not be used in embedded systems!

Function Inlining

It is good programming practice to break out common pieces of computation and accesses to shared data structures into (small) functions. This, however, brings with it the cost of calling a function each time something should be done. In order to mitigate this cost, the compiler transformation of *function inlining* has been developed. Inlining a function means that a copy of the code for the function is placed in the calling function, and the call is removed.

Inlining is a very efficient method to speed up the code, since the function call overhead is avoided but the same computations carried out. Many programmers do this manually by using preprocessor macros for common pieces of code instead of functions, but macros lack the type checking of functions and produce harder-to-find bugs. The executable code will often grow as a result of inlining, since code is being copied into several places.

Inlining may also help shrink the code: for small functions, the code size cost of a function call might be bigger than the code for the function. In this case, inlining a function will actually save code size (as well as speed up the program).

The main problem when inlining for size is to estimate the gains in code size (when optimizing for speed, the gain is almost guaranteed). Since inlining in general increases the code size, the inliner has to be quite conservative. The effect of inlining on code size cannot be exactly determined, since the code of the calling function is disturbed, with non-linear effects.

To reduce the code size, the ideal would be to inline all calls to a function, which allows us to remove the function from the program altogether. This is only possible if all calls are known, i.e. are placed in the same source file as the function, and the function is marked static, so that it cannot be seen from other files. Otherwise, the function will have to be kept (even though it might still be inlined at some calls), and we rely on the linker to remove it if it is not called. Since this decreases the likely gain from inlining, we are less likely to inline such a function.

Low-Level Code Compression

A common transformation on the target code level is to find common sequences of instructions from several functions, and break them out into subroutines. This transformation can be very effective at shrinking the executable code of a program, at the cost of performing more jumps (note that this transformation only introduces machine-level subroutine calls and not full-strength function calls). Experience shows a gain from 10% to 30% for this transformation.

Linker

The linker should be considered an integral part of the compilation system, since there are some transformations that are performed in the linker. The most basic embedded-systems linker should remove all unused functions and variables from a program, and only include the parts of the standard libraries that are actually used. The granularity at which program parts are discarded varies, from files or library modules down to individual functions or even snippets of code. The smaller the granularity, the better the linker. Unfortunately, some linkers derived from desktop systems work on a per-file basis, and this will give unnecessarily big code.

Some linkers also perform post-compilation transformations on the program. Common transformation is the removal of unnecessary bank and page switches (which cannot be done at compile-time since the exact allocation of variable addresses is unknown at that time) and code compression as discussed above extended to the entire program.

Controlling Compiler Optimization

A compiler can be instructed to compile a program with different goals, usually speed or size. For each setting, a set of transformations has been selected that tend to work towards the goal—maximal speed (minimal execution time) or minimal size. The settings should be considered approximate. To give better control, most compilers also allow individual transformations to be enabled or disabled.

For size optimization, the compiler uses a combination of transformations that tend to generate smaller code, but it might fail in some cases, due to the characteristics of the compiled program. As an example, the fact that function inlining is more aggressive for speed optimization makes some programs smaller on the speed setting than on the size setting. The following example data demonstrates this; the two programs were compiled with the same version of the same compiler, using the same memory and data model settings, but optimizing for speed or size:

Program 1	Speed optimization	1301 bytes
	Size optimization	1493 bytes
Program 2	Speed optimization	20432 bytes
	Size optimization	16830 bytes

Program 1 gets slightly smaller with speed optimization, while program 2 is considerably larger, an effect we traced to the fact that function inlining was lucky on program 1. The conclusion is that one should always try to compile a program with different optimization settings and see what happens.

It is often worthwhile to use different compilation settings for different files in a project: put the code that must run very quickly into a separate file and compile that for minimal execution time (maximum speed), and the rest of the code for minimal code size. This will give a small program, which is still fast enough where it matters. Some compilers allow different optimization settings for different functions in the same source file using #pragma directives.

Memory Model

Embedded microcontrollers are usually available in several variants, each with a different amount of program and data memory. For smaller chips, the fact that the amount of memory that can be addressed is limited can be exploited by the compiler to generate smaller code. An

8-bit direct pointer uses less code memory than a 24-bit banked pointer where software has to switch banks before each access. This goes for code as well as data. For example, some Atmel AVR chips have a code area of only 8 kB, which allows a small jump with an offset of +/- 4 kB to reach all code memory, using wrap-around to jump from high addresses to low addresses. Taking advantage of this yields smaller and faster code.

The capacity of the target chip can be communicated to the compiler using a *memory model* option. There are usually several different memory models available, ranging from "small" up to "huge". In general, function calls get more expensive as the amount of code allowed increases, and data accesses and pointers get bigger and more expensive as the amount of accessible data increases. Make sure to use the smallest model that fits your target chip and application—this might give you large savings in code size.

Tips on Programming

The previous section discussed how a compiler works, and gave some examples of how to code better. In this section, we will look at some more concrete tips on how to write compiler-friendly code.

Use the Right Data Size

The semantics of C state that all calculations should have the same result as if all operands were cast to int and the operation performed on int (or unsigned int or long int if values cannot fit in an int). If the result is to be stored in a variable of a smaller type like char, the result is then (conceptually at least) cast down. On any decent 8-bit micro compiler, this process is short-circuited where appropriate, and the entire expression calculated using char or short.

Thus, the size of a data item to be processed should be appropriate for the CPU used. If an unnatural size is chosen, the code generated might get much worse. For example, on an 8-bit micro, accessing and calculating 8-bit data is very efficient. Working with 32-bit values will generate much bigger code and run more slowly, and should only be considered when the data being manipulated need all 32 bits. Using big values also increases the demand for registers for register allocation, since a 32-bit value will require four 8-bit registers to be stored.

On a 32-bit processor, working with smaller data might be inefficient, since the registers are 32 bits. The results of calculations will need to be cast down if the storing variable type is smaller than 32 bits, which introduces shift, mask, and sign-extend operations in the code (depending on how smaller types are represented). On such machines, 32-bit integers should be used for as many variables as possible. chars and shorts should only be used when the precise number of bits are needed (like when doing I/O), or when big types would use too much memory (for example, an array with a large number of elements).

Use the Best Pointer Types

A typical embedded micro has several different pointer types, allowing access to memory in variety of ways, from small zero-page pointers to software-emulated generic pointers. It is obvious that using smaller pointer types is better than using larger pointer types, since both the data space required to store them and the manipulating code is smaller for smaller pointers.

However, there may be several pointers of the same size but with different properties, for example two banked 24-bit pointers huge and far, with the sole difference that huge allows objects to cross bank boundaries. This difference makes the code to manipulate huge pointers much bigger, since each increment or decrement must check for a bank boundary. Unless you really require very large objects, using the smaller pointer variant will save a lot of code space.

For machines with many disjoint memory spaces (like Microchip PIC and Intel 8051), there might be "generic" pointers that can point to all memory spaces. These pointers might be tempting to use, since they are very convenient, but they carry a cost in that special code is needed at each pointer access to check which memory a pointer points to and performing appropriate actions. Also note that using generic pointers typically brings in some library functions (see Section 0).

In summary: use the smallest pointers you can, and avoid any form of generic pointers unless necessary. Remember to check the compiler default pointer type (used for unqualified pointers, and determined by the data memory model used). In many cases it is a rather large pointer type.

Structures and Padding

A C struct is guaranteed to be laid out in memory with the fields in the order of the declaration. However, on processors with alignment restriction on loads and stores, the compiler will probably insert padding between structure members, in order to align each member efficiently. This will make the struct larger than the sum of the sizes of the types of the members, and could break code written under the assumption that structs are laid out contiguously in memory.

C Declaration	Actual Memory Layout

<pre>struct s {
    uint8_t   a;
    uint32_t b;
    uint16_t c;
    int8_t   d;
    int32_t   e;
};</pre>
	<pre>struct s {
	    uint8_t a; /* 1 byte */
	               /* 3 bytes of padding */
	    uint32_t b; /* 4 bytes */
	    uint16_t c; /* 2 bytes */
	    int8_t d; /* 1 byte */
	               /* 1 byte padding */
	    int32_t e; /* 4 bytes */
	};</pre>

Alignment requirements are rare on 8- and 16-bit CPUs, but quite common on 32-bit CPUs. Some CPUs (like the Motorola ColdFire and NEC V850) will generate errors for misaligned loads, while other will only lose performance (Intel x86).

Padding will be inserted at the end of a structure if necessary, to align the size of the structure with the biggest alignment requirement of the machine (typically 4 bytes for a 32-bit machine). This is because every element in an array of structures must start at an aligned boundary in memory. The `sizeof()` operator will reveal the total size of a `struct`, including padding at the end. Incrementing a pointer to a structure will move the pointer `sizeof()` bytes forward in memory, thus reflecting end padding. When a `struct` contains another `struct`, the padding of the member structure is maintained.

In some cases, the compiler offers the ability to pack structures in memory (by `#pragma`, special keywords, or command-line options), removing the padding. This will save data space, but might cost code size, since the code to load misaligned members is potentially much bigger and more complex than the code required to load aligned members.

To make better use of memory, sort the members of the `struct` in order of decreasing size: 32-bit values first, the 16-bit values, and finally 8-bit values. This will make internal padding unnecessary, since each member will be naturally aligned (there will still be padding at end of the `struct` if the size of the `struct` is not an even multiple of the machine word size).

Note that the compiler's padding can break code that uses `struct`s to decode information received over a network or to address memory-mapped I/O areas. This is especially dangerous when code is ported from an architecture without alignment requirements to one with them.

Use Function Prototypes

Function prototypes were introduced in ANSI C as a way to improve type checking. The old style of calling functions without first declaring them was considered unsafe, and is also a hindrance to efficient function calls.

If a function is not properly prototyped, the compiler has to fall back on the language rules dictating that all arguments should be promoted to `int` (or `double`, for floating-point arguments). This means that the function call will be much less efficient, since type casts will have to be inserted to convert the arguments. For a desktop machine, the effect is not very noticeable (most things are the size of `int` or `double` already), but for small embedded systems, the effect is potentially great. Problems include ruining register parameter passing (larger values use more registers) and lots of unnecessary type conversion code.

In many cases, the compiler will give you a warning when a function without a prototype is called. Make sure that no such warnings are present when you compile!

The old way to declare a function before calling it (Kernighan & Ritchie or "K&R" style) was to leave the parameter list empty, like "`extern void foo()`". This is not a proper ANSI prototype and will not help code generation. Unfortunately, few compilers warn about this by default.

The register to parameter assignment for a function can always be inferred from the type of the function, i.e. the complete list of parameter types (as given in the prototype). This means that all calls to a function will use the same registers to store parameters, which is necessary in order to generate correct code. The code in a function does not in any way affect the assignment of registers to parameters.

Use Parameters

As discussed above, register allocation has a hard time with global variables. If you want to improve register allocation, use parameters to pass information to a called function and not shared global variables. Parameters will often be allocated to registers both in the calling and called function, leading to very efficient calls.

Note that the calling conventions of some architectures and compilers limit the number of available registers for parameters, which makes it a good idea to keep the number of parameters down for code that needs to be portable and efficient across a wide range of platforms. It might pay off to split a very complex function into several smaller ones, or to reconsider the data being passed into a function.

Do Not Take Addresses

If you take the address of a local variable (the "&var" construction), it is not likely to be allocated to a register, since it has to have an address and thus a place in memory (usually on the stack). It also has to be written back to memory before each function call, just like a global variable, since some other function might have gotten hold of the address and is expecting the latest value. Taking the address of a global variable does not hurt as much, since they have to have a memory address anyway.

Thus, you should only take the address of a local variable if you really must (it is very seldom necessary). If the taking of addresses is used to receive return values from called functions (from scanf(), for example), introduce a temporary variable to receive the result, and then copy the value from the temporary to the real variable[1]. This should allow the real variable to be register allocated.

Making a global variable static is a good idea (unless it is referred to in another file), since this allows the compiler to know all places where the address is taken, potentially leading to better code.

An example of when not to use the address-of operator is the following, where the use of addresses to access the high byte of a variable will force the variable to the stack. The good way is to use shifts to access parts of values.

Bad example	Good example

`#define highbyte(x) (*((char *)(&x)+1))`
 	
 `short a;`
 `char  b = highbyte(a);`
	`#define highbyte(x) ((x>>8)&0xFF)`
	
	 `short a;`
	 `char  b = highbyte(a);`

---

[1] Note that in C++, reference parameters ("foo(int &)") can introduce pointers to variables in a calling function without the syntax of the call showing that the address of a variable is taken.

Do Not Use Inline Assembly Language

Using inline assembly is a very efficient way of hampering the compiler's optimizer. Since there is a block of code that the compiler knows nothing about, it cannot optimize across that block. In many cases, variables will be forced to memory and most optimizations turned off.

The output of a function containing inline assembly should be inspected after each compilation run to make sure that the assembly code still works as intended. In addition, the portability of inline assembly is very poor, both across machines (obviously) and across different compilers for the same target.

If you need to use assembler, the best solution is to split it out into assembly source files, or at least into functions containing only inline assembly. Do not mix C code and assembly code in the same function!

Do Not Write Clever Code

Some C programmers believe that writing fewer source code characters and making clever use of C constructions will make the code smaller or faster. The result is code which is harder to read, and which is also harder to compile. Writing things in a straightforward way helps both humans and compilers understand your code, giving you better results. For example, conditional expressions gain from being clearly expressed as conditions.

Consider the two ways to set the lowest bit of variable b if the lower 21 bits of another (32-bit) variable are non-zero as illustrated below. The clever code uses the! operator in C, which returns zero if the argument is non-zero ("true" in C is any value except zero), and one if the argument is zero.

The straightforward solution is easy to compile into a conditional followed by a set bit instruction, since the bit-setting operation is obvious and the masking is likely to be more efficient than the shift. Ideally, the two solutions should generate the same code. The clever code, however, may result in more code since it performs two ! operations, each of which may be compiled into a conditional.

"Clever" solution	Straightforward solution

<pre>unsigned long int a;
unsigned char   b;
 	
/* Move bits 0..20 to positions 11..31
 * If non-zero, first ! gives 0   */
b |= !!(a << 11);</pre>
	<pre>unsigned long int a;
	unsigned char b;
	
	/* Straight-forward if statement */
	if((a & 0x1FFFFF) != 0)
	    b |= 0x01;</pre>

Another example is the use of conditional values in calculations. The "clever" code will result in larger machine code, since the generated code will contain the same test as the straightforward code, and adds a temporary variable to hold the one or zero to add to str. The straightforward code can use a simple increment operation rather than a full addition, and does not require the generation of any intermediate results.

"Clever" solution	Straightforward solution

```int bar(char *str)```
```{```
  ```/* Calculating with result of */```
  ```/* comparison.   */```
  ```return foo(str+(*str=='+'));```
```}```
	```int bar(char *str)```
	```{```
	  ```if(*str=='+')```
	    ```str++;```
	  ```return foo(str);```
	```}```

Since clever code almost never compiles better than straightforward code, why write clever code? From a maintenance standpoint, writing simpler and more understandable code is definitely the method of choice.

Use Switch for Jump Tables

If you want a jump table, see if you can use a switch statement to achieve the same effect. It is quite likely that the compiler will generate better and smaller code for the switch rather than a series of indirect function calls through a table. Also, using the switch makes the program flow explicit, helping the compiler optimize the surrounding code better. It is very likely that the compiler will generate a jump table, at least for a small dense switch (where all or most values are used).

Using a switch is also more reliable across machines; the layout that may be optimal on one CPU may not be optimal on another, but the compiler for each will know how to make the best possible jump table for both. The switch statement was put into the C language to facilitate multiway jumps: use it!

Investigate Bit Fields Before Using Them

Bitfields offer a very readable way to address small groups of bits as integers, but the bit layout is implementation defined, which creates problems for portable code. The code generated for bitfields will be of very varying quality, since not all compilers consider them very important. Some compilers will generate incredibly poor code since they do not consider them worth optimizing, while others will optimize the operations so that the code is as efficient as manual masking and shifting.

The advice is to test a few bitfield variables and check that the bit layout is as expected, and that the operations are efficiently implemented. If several compilers are being used, check that they all have the same bit layout. In general, using explicit masks and shifts will generate more reliable code across more targets and compilers.

Watch Out for Library Functions

As discussed above, the linker has to bring in all library functions used by a program with the program. This is obvious for C standard library functions like printf() and strcat(), but there are also large parts of the library which are brought in implicitly when certain types of arithmetic are needed, most notably floating point. Due to the way in which C performs implicit type conversions inside expressions, it is quite easy to inadvertently bring in floating point, even if no floating point variables are being used.

For example, the following code will bring in floating point, since the `ImportantRatio` constant is of floating point type—even if its value would be `1.95*20==39`, and all variables are integers:

Example of Accidental Floating Point

`#define ImportantRatio (1.95*Other)`
`... int`
`temp = a * b + CONSTANT * ImportantRatio;`

If a small change to a program causes a big change in program size, look at the library functions included after linking. Especially floating point and 32-bit integer libraries can be insidious, and creep in due to C implicit casts.

Another way to shrink the code of your program is to use limited versions of standard functions. For instance, the standard `printf()` is a very big function. Unless you really need the full functionality, you should use a limited version that only handles basic formatting or ignores floating point. Note that this should be done at link time: the source code is the same, but a simpler version is linked. Because the first argument to `printf()` is a string, and can be provided as a variable, it is not possible for the compiler to automatically figure out which parts of the function your program needs.

Use Extra Hints

Some compilers allow the programmer to specify useful information that the compiler cannot deduce itself to help optimize the code. For example, DSP compilers often allow users to specify that two pointer or array arguments are unaliased, which helps the compiler optimize code accessing these two arrays simultaneously. Other examples are the specification of function as *pure* (without side effects) or *tasks* (will loop forever, thus no need to save registers on entry). A common example is *inline*, which might be considered a hint or an order by the compiler. This information is usually introduced using non-portable keywords and should be put in tuned header files (if possible). It might give great benefits in code efficiency, however.

Final Notes

This chapter has tried to give you an idea of how a modern C compiler works, and to give you some concrete hints on how you can get smaller code by using the compiler wisely. A compiler is a very complex system with highly non-linear behavior, where a seemingly small change in the source code can have big effects on the assembly code generated.

The basis for the compilation process is that the compiler should be able to understand what your code is supposed to do, in order to perform the operations in the best possible way for a given target. As a general rule, code that is easy to understand for a fellow human programmer—and thus easy to maintain and port—is also easier to compile efficiently.

Note that unless you let your compiler use its higher optimization levels, you have wasted a lot of your investment. What you pay for when you buy a compiler is mostly the work put into developing and tuning optimizations for a certain target, and if you do not use these optimizations, you are not using your compiler to its best effect.

Choose your compiler wisely: different compilers for the same chip can be very different. Some are better at generating fast code, other at generating small code, and some may be no good at all. To evaluate a compiler, the best way is to use a demo version to compile small portions of your own "typical" code. Some chip vendors also provide benchmark tests of various compilers for their chips, usually targeted towards the intended application area for their chips. The compiler vendor's own benchmarks should be taken with some skepticism; it is (almost) always possible to find a program where a certain compiler performs better than the competition.

For more tips on efficient C programming for embedded systems, you should check out the classes presented at embedded systems trade shows (around the world). The websites of the companies making compilers often contain some technical notes or white papers on particular tricks for their compilers (but make sure to watch out for those that are pure marketing material!). In particular, I would like to recommend the IAR web site section with white papers and articles: http://www.iar.com/Press/Articles.

Acknowledgements

This text was developed with much help from the compiler developers at IAR Systems, and most of the work was performed during my time at IAR. I had some really fun and educating years at IAR.

Optimizing C and C++ Code

Sandeep Ahluwalia

Sandeep Ahluwalia is a writer and developer for EventHelix.com Inc., a company that is dedicated to developing tools and techniques for real-time and embedded systems. They are developers of the EventStudio, a CASE tool for distributed system design in object oriented as well as structured development environments. They are based in Gaithersburg, Maryland. Their website is www.eventhelix.com

Embedded software often runs on processors with limited computation power, so optimizing the code becomes a necessity. In this chapter we will explore the following optimization techniques for C and C++ code developed for real-time and embedded systems.

Adjust Structure Sizes to Power of Two

When arrays of structures are involved, the compiler performs a multiply by the structure size to perform the array indexing. If the structure size is a power of 2, an expensive multiply operation will be replaced by an inexpensive shift operation. Thus, keeping structure sizes aligned to a power of 2 will improve performance in array indexing.

Place Case Labels in Narrow Range

If the case labels are in a narrow range, the compiler does not generate a if-else-if cascade for the switch statement. Instead, it generates a jump table of case labels along with manipulating the value of the switch to index the table. This code generated is faster than if-else-if cascade code that is generated in cases where the case labels are far apart. Also, performance of a jump table based switch statement is independent of the number of case entries in the switch statement.

Place Frequent Case Labels First

If the case labels are placed far apart, the compiler will generate if-else-if cascaded code with comparing for each case label and jumping to the action for leg on hitting a label match. By placing the frequent case labels first, you can reduce the number of comparisons that will be performed for frequently occurring scenarios. Typically this means that cases corresponding to the success of an operation should be placed before cases of failure handling.

Break Big Switch Statements into Nested Switches

The previous technique does not work for some compilers as they do not generate the cascade of if-else-if in the order specified in the switch statement. In such cases nested switch statements can be used to get the same effect.

To reduce the number of comparisons being performed, judiciously break big switch statements into nested switches. Put frequently occurring case labels into one switch and keep the rest of case labels in another switch which is the default leg of the first switch.

Splitting a Switch Statement

```
// This switch statement performs a switch on frequent messages and handles the
// infrequent messages with another switch statement in the default leg of the outer

// switch statement
pMsg = ReceiveMessage();
switch (pMsg->type)
{
case FREQUENT_MSG1:
    handleFrequentMsg1();
    break;

case FREQUENT_MSG2:
    handleFrequentMsg2();
    break;

. . .

case FREQUENT_MSGn:
    handleFrequentMsgn();
    break;

default:
    // Nested switch statement for handling infrequent messages.
    switch (pMsg->type)
    {
    case INFREQUENT_MSG1:
        handleInfrequentMsg1();
        break;

case INFREQUENT_MSG2:
    handleInfrequentMsg2();
    break;

. . .

case INFREQUENT_MSGm:
    handleInfrequentMsgm();
    break;
    }
}
```

Minimize Local Variables

If the number of local variables in a function is small, the compiler will be able to fit them into registers. Hence, it will be avoiding frame pointer operations on local variables that are kept on the stack. This can result in considerable improvement for two reasons:

- All local variables are in registers so this improves performance over accessing them from memory.
- If no local variables need to be saved on the stack, the compiler will not incur the overhead of setting up and restoring the frame pointer.

Declare Local Variables in the Innermost Scope

Do not declare all the local variables in the outermost function scope. You will get better performance if local variables are declared in the innermost scope. Consider the example below; here object a is needed only in the error case, so it should be invoked only inside the error check. If this parameter was declared in the outermost scope, all function calls would have incurred the overhead of object a's creation (i.e., invoking the default constructor for a).

```
                    Local variable scope

int foo(char *pName)
{
    if (pName == NULL)
    {
        A a;
        ...
        return ERROR;
    }
    ...
    return SUCCESS;
}
```

Reduce the Number of Parameters

Function calls with large numbers of parameters may be expensive due to a large number of parameter pushes on stack on each call. For the same reason, avoid passing complete structures as parameters. Use pointers and references in such cases.

Use References for Parameter Passing and Return Value for Types Bigger than 4 Bytes

Passing parameters by value results in the complete parameter being copied on to the stack. This is fine for regular types like integer, pointer etc. These types are generally restricted to four bytes. When passing bigger types, the cost of copying the object on the stack can be prohibitive. In the case of classes there will be an additional overhead of invoking the constructor for the temporary copy that is created on the stack. When the function exits the destructor will also be invoked.

Thus, it is efficient to pass references as parameters. This way you save on the overhead of a temporary object creation, copying and destruction. This optimization can be performed easily without a major impact to the code by replacing pass by value parameters with const references. (It is important to pass const references so that a bug in the called function does not change the actual value of the parameter.)

Passing bigger objects as return values also has the same performance issues. A temporary return object is created in this case too.

Don't Define a Return Value if Not Used

The called function does not "know" if the return value is being used. So, it will always pass the return value. This return value passing may be avoided by not defining a return value which is not being used.

Consider Locality of Reference for Code and Data

The processor keeps data or code that is referenced in cache so that on its next reference it gets it from cache. These cache references are faster. Hence it is recommended that code and data that are being used together should actually be placed together physically. This is actually enforced into the language in C++. In C++, all the object's data is in one place and so is code. When coding is C, the declaration order of related code and functions can be arranged so that closely coupled code and data are declared together.

Prefer int over char and short

With C and C++ prefer use of int over char and short. The main reason behind this is that C and C++ perform arithmetic operations and parameter passing at the integer level. If you have an integer value that can fit in a byte, you should still consider using an int to hold the number. If you use a char, the compiler will first convert the values into integer, perform the operations and then convert back the result to char.

Let's consider the following code which presents two functions that perform the same operation with char and int.

Comparing char and int operations
| ```
char sum_char(char a, char b)
{
    char c;
    c = a + b;
    return c;
}

int sum_int(int a, int b)
{
    int c;
    c = a + b;
    return c;
}
``` |
```

A call to `sum_char` involves the following operations:

1. Convert the second parameter into an `int` by sign extension (C and C++ push parameters in reverse).

2. Push the sign extended parameter on the stack as b.

3. Convert the first parameter into an `int` by sign extension.

4. Push the sign extended parameter on to the stack as a.

5. The called function adds a and b.

6. The result is cast to a `char`.

7. The result is stored in `char` c.

8. c is again sign extended.

9. Sign extended c is copied into the return value register and function returns to caller.

10. The caller now converts again from `int` to `char`.

11. The result is stored.

A call to `sum_int` involves the following operations:

1. Push `int` b on stack

2. Push `int` a on stack

3. Called function adds a and b

4. Result is stored in `int` c

5. c is copied into the return value register and function returns to caller.

6. The called function stores the returned value.

Thus we can conclude that `int` should be used for all integer variables unless storage requirements force us to use a `char` or `short`. When `char` and `short` have to be used, consider the impact of byte alignment and ordering to see if you would really save space. (Many processors align structure elements at 16 byte boundaries)

Define Lightweight Constructors

As far as possible, keep the constructor lightweight. The constructor will be invoked for every object creation. Keep in mind that many times the compiler might be creating temporary objects over and above the explicit object creations in your program. Thus, optimizing the constructor might give you a big boost in performance. If you have an array of objects, the default constructor for the object should be optimized first as the constructor gets invoked for every object in the array.

Prefer Initialization Over Assignment

Consider the following example of a complex number:

Initialization and assignment
| ```
void foo()
{
    Complex c;
    c = (Complex)5;
}
void foo_optimized()
{
    Complex c = 5;
}
``` |

In the function foo, the complex number c is being initialized first by the instantiation and then by the assignment. In foo_optimized, c is being initialized directly to the final value, thus saving a call to the default constructor of Complex.

Use Constructor Initialization Lists

Use constructor initialization lists to initialize the embedded variables to the final initialization values. Assignments within the constructor body will result in lower performance as the default constructor for the embedded objects would have been invoked anyway. Using constructor initialization lists will directly result in invoking the right constructor, thus saving the overhead of default constructor invocation.

In the following example, the optimized version of the Employee constructor saves the default constructor calls for m_name and m_designation strings.

Constructor initialization lists
| ```
Employee::Employee(String name, String designation)
{
    m_name = name;
    m_designation = designation;
}

/* === Optimized Version === */

Employee::Employee(String name, String designation): m_name(name),
m_destignation (designation)
{
}
``` |

Do Not Declare "Just in Case" Virtual Functions

Virtual function calls are more expensive than regular function calls, so do not make functions virtual "just in case" somebody needs to override the default behavior. If the need arises, the developer can just as well edit the additional base class header file to change the declaration to virtual.

Inline 1 to 3 Line Functions

Converting small functions (1 to 3 lines) into inline will give you big improvements in throughput. Inlining will remove the overhead of a function call and associated parameter passing. But using this technique for bigger functions can have negative impact on performance due to the associated code bloat. Also keep in mind that making a method inline should not increase the dependencies by requiring a explicit header file inclusion when you could have managed by just using a forward reference in the non-inline version. (See the chapter on header file include patterns for more details.)

Do Not Declare "Just in Case" Virtual Functions

Inline 1 to 3 Line Functions

Real-Time Asserts

Jack Ganssle

Expert programmers use the `assert()` macro to seed their code with debugging statements. So should we all.

Niall Murphy wrote a pair of articles in the April and May 2001 issues *of Embedded Systems Programming*. Niall does a great job of describing the what, how and why of this debugging construct. Do check them out. Basically, if `assert()`'s argument is non-zero it flags an error message.

The macro must have no effect on the system—it shouldn't change memory, interrupt state, or I/O, and it should make no function calls. The user must feel free to sprinkle these at will throughout the program. Disabling the asserts in released code must never change the way the program operates.

The macro generally looks something like:

```
#define assert(s)
    if (s)
        {}
    else
        printf("Error at %s %d: ", __FILE__, __LINE__)
```

It's sort of an odd way to write code—if the condition is true (in which case `assert()` takes no action) there's a null pair of braces. Why not test the negation of the statement "`s`"?

The answer lies in my, uh, assertion above: the macro must have no effect on the system. Code it without the `else` and poorly written software, perhaps including an unmatched `else` from another `if` (this is a debugging construct after all; it's role is to look for dumb mistakes) will cause the assert to change the system's behavior in unpredictable ways.

Embedded Issues

Few embedded systems make use of the `assert()` macro since we're usually in an environment with neither display console nor `printf` statement. Where do the error messages go? Equally problematic, embedding strings ("Error at" and the file name) eats valuable ROM.

Replace the `printf` with a software interrupt. Plenty of processors have a one byte or single word instruction that vectors the execution stream to an interrupt handler. Write code to capture the exception and take some sort of action. Log the error to a data structure, flash an LED, or put the device in a safe mode.

Initialize your debugging tools to automatically set a breakpoint on the handler. You'll forget the breakpoint is set. But once in a while, Surprise! The system suddenly stops, the debug window highlighting the first line of the exception routine. You've captured a bug in its early phases, when it's easiest to track.

Systems constrained by small ROM sizes can leave out the __LINE__ and __FILE__ arguments. Pop the stack and log the caller's return address. It's an easy matter to track back to find which assertion threw the exception.

Most developers embed assert() inside a preprocessor conditional. Flip a compile-time switch and they all disappear. This gives us a highly-debuggable version of the code for our own use, and a production release not seeded with the macros.

But I'm not keen on shipping code that's differs from our test version. NASA's commandment to "test what you fly and fly what you test" makes an awful lot of sense for building spacecraft as well as any embedded system. When ROM and CPU cycles aren't in short supply leave the assert()s seeded in the code, enabled, not #defined out. If you believe the system can merrily proceed despite an assertion failure, code the exception handler to log the error, leaving debugging breadcrumbs behind, and return.

Is your system connected to the 'net? Why not send an email message logging the error with a snapshot of the snapshot to crash@yourcompany.com? Be sure to include the firmware's version number.

This is yet another reason to get and use a version control system. If your code isn't static, if new versions proliferate like politician's promises of tax relief, the stack contents will be meaningless unless you can recreate earlier releases.

Embedded systems should use asserts to check housekeeping problems, like stack growth. On an x86 processor the instruction mov ax,sp grabs the current stack pointer. Embed this, or a similar instruction for your CPU, into an assert that then compares against a nearly full stack. Seed every task with the macro.

Check malloc's return value, too, if you're gutsy enough to use it in firmware. If malloc fails it always returns an error code. Ignore it at your peril!

Real-Time Asserts

When designing the system, answer two questions: how fast is fast enough? How will you know if you've reached this goal?

Some people are born lucky. Not me. I've learned that nature is perverse, and will get me if it can. Call it high-tech paranoia. Plan for problems, and develop solutions for those problems before they occur. Assume each ISR will be too slow, and plan accordingly.

I'm frustrated that so many developers of real-time systems think only in the procedural domain, working diligently to build a sequential series of instructions that implement some algorithm, blithely ignoring the time aspects of their code. Building real-time code means managing time as seriously as we build if-elses. Track free CPU microseconds as well as available bytes of ROM.

There's surprisingly little interest in managing time. C and C++ compilers give us no help. They should produce a listing which gives us the execution time, or at least a range of times, for each statement in our program. We compiler users have no idea if a simple ++int takes a microsecond or a week. More complex statements invoke the compiler's runtime library, truly a black hole of unknown execution times.

Years ago a friend worked on a Navy job using the CMS-2 language. The compiler was so bug-ridden that they adopted a unique coding strategy: write a function, compile it, and see if the assembly output made sense. If clearly wrong, change something in the CMS-2 source— even spacing—and hope a recompile produced better results. In effect, they made random changes to their source in hopes of taming the compiler.

We do the same thing when the compiler does not produce timing information. How do we fix a slow snippet? Change something—almost anything—and hope. That's more akin to hacking than to software engineering. Seems to me the vendors do their customers a horrible disservice by not documenting best, worst and average execution times.

A very cool product, despite its high price, was Applied Microsystem's CodeTest. Among other things, it measured execution time of each function in a program, seeding the code with special instructions whose action was captured by a hardware probe. Yet Applied failed, never able to generate much interest in that product. MetroWerks (www.metrowerks.com) sells the device now; I wish them success, and hope more of the embedded community seeks tools for managing the time part of their applications.

In lieu of buying time tools, why not build real-time assert()s, macros that help us debug in the time domain?

The oldest timing trick is to drive an output bit high when entering a routine, and then low as we exit. Probe the port with an oscilloscope and measure the time the bit is set. Whether you're working on an 8-bit microcontroller or a 32-bit VME-based system, always dedicate one or two parallel I/O bits to debugging. That is, *have the hardware designers include a couple of output bits just for software debugging purposes.* The cost is vanishingly small; the benefits often profound.

Rather than dropping a pair of outport() C statements directly into the code, I prefer to define a assert-like macros. Never sprinkle low level I/O into anything other than drivers, except in the most ROM-constrained systems.

Call them RT_set_assert() and RT_unset_assert(), the RT meaning real-time to clearly indicate that these are not your father's asserts anymore. Like this:

```
ISR_entry:
    push all registers
    rt_set_assert()
    service interrupt
    rt_unset_assert()
    pop registers
    return
```

The scope might show:

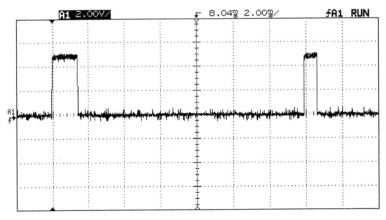

Figure 20-1.

In this example we see two invocations of the ISR. The first time (note the time base setting is 2 msec/division) the routine runs for about 2.5 msec. Next time (presumably the routine includes conditional code) it runs for .06 msec.

We also clearly see a 14 msec period between executions. If these two samples are indicative of the system's typical operation, the total CPU overhead dedicated to this one interrupt is (2.5 msec + 0.06 msec)/14 msec, or 22%.

Crank up the scope's time base and you can measure the ISR's execution time to any desired level of precision.

This brain-dead trick has been around forever. But you can learn a lot about the behavior of a system by extending the idea just a bit. Want to measure min and max execution time of a function or ISR? Set a digital scope to store multiple acquisitions on the screen. Trigger on the I/O bit's positive edge. You'll see the bit go high, a clear space on the screen, and then a lot of hash before it goes low. The clear space is the minimum time. The interval from the positive edge to the end of the hash is the maximum execution time.

I often want to know the total amount of time spent servicing interrupts. Heavy interrupt loads are a sure sign of a system which will be a bear to debug and maintain. Drop the RT_assert pair into every ISR. That is, every ISR toggles a common bit. If the scope trace stays mostly low, there's little loading. Mostly high, and kiss your family goodbye for the duration of the project.

Perhaps an even more profound measurement is the system's total idle time. Is the CPU 100% loaded? 90%? Without this knowledge you cannot reliably tell the boss "sure, we can add that feature."

Instead of driving the debug bit in ISRs, toggle it in the idle loop. Applications based on RTOSes often don't use idle loops, so create a low priority idle task that runs when there's nothing to do.

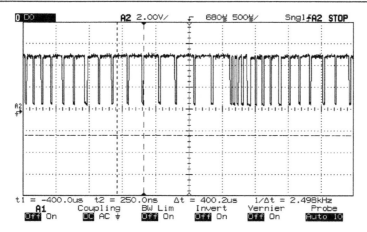

Figure 20-2: Interrupt loading—this puppy is way too busy servicing interrupts.

The instrumented idle loop looks like:

```
idle:
    rt_set_assert()
    rt_unset_assert()low
    look for something to do
    jump to idle
```

While the idle loop runs the debug bit wiggles like mad. If you turn the scope's time base down (to more time per division) the toggling bit looks more like hash, with long down periods indicating that the code is no longer in the idle loop. In the next figure about a third of the processing time is unused.

If an interrupt occurs after setting the bit high, but before returning it to zero, then the "busy" interval will look like a one on the scope and not a zero. Idle times are those where you see hash—the signal rapidly cycling up and down. Busy times are those where the signal is a steady one or zero.

We developers are being shortchanged by our tools; the continued disappearance of the in-circuit emulator gives us ever less insight into our projects' time domains. But we're lucky in that most high integration CPUs include plenty of timers. Even simple systems today use RTOSes sequenced by interrupts from a timer. If you're not using an RTOS, do program one of these peripherals to track time for the sake of performance measurements.

Create a simple RT_assert pair to compute elapsed time over a function or region of code. `RT_enter_assert()` saves the clock; `RT_exit_assert()` figures time, in ticks or microseconds. In a preemptive system it's best to pass an argument with a unique identifier for each code snippet being measured. This is then an index into a data structure, so that multiple instances of the assert don't clobber each other's data.

In a real system logging every instantiation of the code chunk would generate a flood of data. Instead, have `RT_exit_assert()` compute the max execution time. Or, if needed, figure min and max.

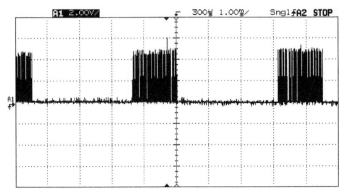

Figure 20-3.

Examine the structure with a debugger to get a quick code profile.

How much accuracy do you need? If the timer is used for context switching, no doubt it only generates an interrupt every handful of milliseconds. Yet the hardware actually counts CPU clocks, or CPU clocks divided by some very small number. The RT_assert macros can read OS-maintained time (which counts interrupts) and the timer's hardware. Resolution to a few microseconds is possible.

To "tare" the readings, compensating for the overhead contributed by the RT_asserts, issue the two macros back to back. Subtract the tare value to increase precision.

Multitasking can queer the results. Preempting the code in the middle of a profiling operation then includes the execution time of the preempter. Just as we want to include the overhead contributed by the compiler's runtime routines, most of the time we want to know elapsed time, which includes preemption. Alternatively, disable interrupts during the measurement, though of course that may greatly affect system operation.

Some developers advocate using macros to send unique codes, a start/stop pair per function, to a parallel port. A logic analyzer then acquires the data. This is more trouble than it's worth. The analyzer quickly accumulates thousands or millions of data points; sometimes much more than it can store. It's impossible to make sense of the information without writing a computer program to empty the tool's buffer and convert raw codes and time into meaningful info.

Worse, the analyzer collects data only during the debugging phase, and only then when you're specifically taking performance measurements. I believe in proactive debugging—seed the code and the execution environment with tools and tricks to log data and flag errors *all* the time.

And that's where the idea of "real time asserts" shines. Expand `RT_exit_assert()` just a bit. Pass a parameter that indicates the maximum allowed time. The macro computes the actual period… and errors if it exceeds what we've expected. Maybe you leave this in only during debugging to grab your attention when the system snarls. Perhaps it's deployed with the production release to take some action when insane bottlenecks occur.

After all, missing a deadline is as much of a bug in many real-time systems as is a complete system crash. If we use asserts to capture errors in the procedural domain, we should do no less in the time domain.

Errors and Changes

Introduction

Jack Ganssle

In this section we look at error handling, error management, managing changes to memory, and instrumenting code to detect errors early.

Two-thirds of all system crashes come from unmanaged, or poorly managed, exceptions. We rely on testing to prove our systems correct, yet testing doesn't work well. Some studies show that testing exercises about half the code—just half. Much of the untested software is in the exception handlers, which are notoriously difficult to check.

Obviously—or obviously to me—we've got to take strong action to deal with these errors. Alan Griffiths shows how to deal with exceptions in the C++ environment. I write about watchdog timers and debugging proactively.

In 2003 a woman almost died when her rice cooker reprogrammed her pacemaker's Flash memory. A man collapsed when a retail store's security device did the same to his. Bill Gatliff tells us how to download code to Flash *safely* … we better use his advice.

You have to think through error handling from the outset of designing a product. Can the system recover from a detected error? How?

Will you permit a system crash or at least leave some debugging breadcrumbs behind? How?

Does a longjump back to main() suitably reset the system after a catastrophic error is detected? Are you sure the peripherals will reset properly? Maybe it makes sense to halt and let the watchdog time out.

But above all, error handling is fundamental to building a reliable embedded system.

Implementing Downloadable Firmware With Flash Memory
A Microprogrammer-based approach to firmware updates
William Gatliff

Bill Gatliff is a freelance embedded developer and training consultant with almost ten years of experience using GNU and other tools for building embedded systems. His product background includes automotive, industrial, aerospace and medical instrumentation applications.

Bill specializes in GNU-based embedded development, and in using and adapting GNU tools to meet the needs of difficult development problems. He welcomes the opportunity to participate in projects of all types.

Bill is a Contributing Editor for *Embedded Systems Programming* (http://www.embedded.com/), a member of the Advisory Panel for the Embedded Systems Conference (http://www.esconline.com/), maintainer of the Crossgcc FAQ, creator of the gdbstubs (http://sourceforge.net/projects/gdbstubs) project, and a noted author and speaker.

Bill welcomes feedback and suggestions. Contact information is on his website, at http://www.billgatliff.com.

Introduction

The problem with any approach to *in-situ* firmware updates is that when such a feature contains a flaw, the target system may become an expensive doorstop—and perhaps injure the user in the process. Many of the potential pitfalls are obvious and straightforward to correct, but other, insidious defects may not appear until after a product has been deployed in its application environment.

Users are unequaled in their abilities to expose and exploit product defects, and to make matters worse, users also generally fail to heed warnings like "system damage will occur if power is interrupted while programming is underway." They will happily attempt to reboot an otherwise functional system in the middle of the update process, and then file a warranty claim for the now "defective" product.

Any well-designed, user-ready embedded system must include the ability to recover from user errors and other catastrophic events to the fullest extent possible. The best way to accomplish this is to implement a fundamentally sound firmware update strategy that avoids these problems entirely. This chapter presents one such design.

The Microprogrammer

The following sections describe a *microprogrammer*-based approach to the implementation of a downloadable firmware feature for embedded systems. This approach is suitable for direct implementation as described, but can also be modified for use in situations where some of its features are not needed, or must be avoided.

The definition of a *microprogrammer* is a system-level description of how the embedded system behaves before, during and after the firmware update process. This behavior is carefully defined to help avoid many of the problems associated with other approaches to downloadable firmware. A careful implementation of this behavior eliminates the remaining concerns.

The first step in a microprogrammer-based firmware update process is to place the embedded system into a state where it is expecting the download process to occur. The transition to this state could be caused by the user pressing a button marked **UPGRADE** on the device's interface panel, by the system detecting the start or end of a file transfer, or by some other means. In any case, the target system now realizes that its firmware will soon be updated, and brings any controlled processes to a safe and stable halt configuration.

Next, the target system is sent a small application called a *microprogrammer*. The microprogrammer assumes control of the system, and begins receiving the new application firmware and programming it into flash. At the conclusion of this process, the target system begins running the new firmware.

Advantages of Microprogrammers

One of the biggest selling points of a microprogrammer-based approach to downloadable firmware is its flexibility. The microprogrammer's implementation can be changed even in the final moments before the firmware update process begins, which allows bug fixes and enhancements to be applied retroactively to deployed systems.

A microprogrammer does not consume resources in the target system except when programming is actually underway. Furthermore, since an effective microprogrammer can be as small as 10K of code or less, the target system does not require a bulky, sophisticated communications protocol in order to receive the microprogrammer at the start of the process— a simple text file transfer is often reliable enough.

The safety of a microprogrammer-based firmware update process is sometimes its most compelling advantage. When the target system's flash chip is not used for any purpose other than firmware storage, the code needed to erase and program the flash chip *does not exist* in the system until it is actually needed. As such, the system is highly unlikely to accidentally erase its own firmware, even in severe cases of program runaway.

Disadvantages of Microprogrammers

One of the drawbacks of microprogrammers is that the microprogrammer itself is usually implemented as a standalone program, which means that its code is managed separately from both the application that downloads it to the target system, and the application that the microprogrammer delivers to the target system. This management effort requires additional developer resources, the quantity of which is highly dependent upon how closely coupled these applications are to each other. Careful attention to program modularity helps to minimize the workload.

A microprogrammer is generally downloaded to and run from RAM, which means that the target system needs some quantity of memory available to hold the microprogrammer. This memory can be shared with the preexisting target application when necessary, but an embedded system that only has a few hundred bytes of RAM in total will probably need a different strategy.

And finally, in a single-flash system the microprogrammer approach requires the target system to be able to run code from RAM. This simply isn't possible for certain microprocessor architectures, in particular the venerable 8051 and family. Hardware-oriented workarounds exist for these cases, but the limitations they impose often outweigh their benefits.

Receiving a Microprogrammer

The code in Listing 21-1 illustrates the functionality needed by a target system to download and run a microprogrammer. In the example, the target is sent a plaintext Motorola S Record file from some I/O channel (perhaps a serial port), which is decoded and written to RAM. The target then activates the microprogrammer by jumping into the downloaded code at the end of the transfer.

Notice that the `programmer_buf[]` memory space is allocated as an automatic variable, which means that it has no fixed location in the target system's memory image. This implies both that the addresses in the incoming S Records are relative rather than absolute, and that the incoming code is position-independent. If your compiler cannot produce position-independent code, then `programmer_buf[]` must be assigned to a fixed location in memory and the addresses in the incoming S Records must be located within that memory space.

The incoming microprogrammer image can be placed on top of other data if the target system does not have the resources to permanently allocate `programmer_buf[]`. At this point the embedded system has suspended normal operations anyway, making the bulk of its total RAM space available for use by the microprogrammer.

```
enum srec_type_t {
    SREC_TYPE_S1, SREC_TYPE_S2, SREC_TYPE_S3,
    SREC_TYPE_S7, SREC_TYPE_S9
};
typedef void (*entrypoint_t)(void);

void microprogrammer()
{
    char programmer_buf[8192];
    int len;
    char sbuf[256];
    unsigned long addr;
    enum srec_type_t type;
    entrypoint_t entrypoint;

    while (1) {
        if (read_srecord(&type, &len, &addr, sbuf)) {
            switch (type) {
                case SREC_TYPE_S1:
                case SREC_TYPE_S2:
                case SREC_TYPE_S3:
                    /* record contains data (code) */
                    memcpy(programmer_buf + addr, sbuf, len);
                    break;

                case SREC_TYPE_S7:
                    /* record contains address of downloaded main() */
                    entrypoint = (entrypoint_t)(programmer_buf + addr);
                    break;

                case SREC_TYPE_S9:
                    /* record indicates end of data (code)- run it */
                    entrypoint();
                    break;
            }
        }
    }
}
```

Listing 21-1: Code that downloads and runs a microprogrammer.

A Basic Microprogrammer

The top-level code for a microprogrammer is shown in Listing 21-2. For consistency with the previous example, this code also receives an S Record file from some source and decodes it. The microprogrammer writes the incoming data to flash, and the system is rebooted at the end of the file transfer. Although overly simplistic (a plaintext file transfer is probably not reliable enough for large programs), this code illustrates all the important features of a microprogrammer.

```
void programmer()
{
    int len;
    char buf[256];
    unsigned long addr;
    enum srec_type_t type;

    while (1) {
        if (read_srecord(&type, &len, &addr, buf)) {
            switch (type) {
                case SREC_TYPE_S1:
                case SREC_TYPE_S2:
                case SREC_TYPE_S3:
                    /* record contains data or code- program it */
                    if (!is_section_erased(addr, len))
                        erase_flash(addr, len);
                    write_flash(addr, len, buf);
                    break;

                case SREC_TYPE_S9:
                    /* this record indicates end of data-
                        execute system reset to run new application */
                    reset();
                    break;
            }
        }
    }
}
```

Listing 21-2: A basic microprogrammer.

In addition to actually erasing flash, the function `erase_flash()` also manages a simple data structure that keeps track of which sections of the flash chip need to be erased, and which ones have already been erased. This data structure is checked by the `is_section_erased()` function, which prevents multiple erasures of flash sections when data arrives out of order—which is a common occurrence in an S Record file.

Common Problems and Their Solutions

Regardless of how you modify the microprogrammer-based system description to fit your own requirements, you will encounter some common problems in the final implementation. These problems, along with their solutions, are the subject of this section.

Debugger Doesn't Like writeable code Space

Some debuggers, emulators in particular, struggle with the idea of code space that can be written to by the target microprocessor. Most debugging tools treat code space as read-only, and some will generate error messages or simply disallow a writing operation when they detect one.

In general, the debugger's efforts to protect code space are well-intentioned. A program writing to its code space usually signals a serious programming error, *except* when a firmware update is in progress. The debugger can't tell the difference, of course. The remedy is to implement a *memory alias* for the flash chip, in a different memory space that is not considered read-only by the debugger.

Consider a 512KB flash chip that starts at address 0 in the target's memory space. By utilizing a chip select line that responds to an address in the range of 0-1024KB, you can access the first byte in the flash chip by writing to address 0x80000 instead of address 0, and simultaneously avoid any intrusion from the debugger or emulator.

The memory region between 512KB and 1024KB is sometimes called an *alias* of the region 0-512KB, because the underlying physical hardware cannot distinguish between the two addresses and thus maps them to the same physical location in the flash chip. The debugger *can* distinguish between the two address ranges, however, and can therefore be configured to ignore (and thereby permit) write accesses in the alias region.

The typical implementation of a memory alias is straightforward: simply double the size of the chip select used to activate the device, and apply an offset to any address that refers to a location in the region designated as the "physical" address region, to move it to the "alias" region. The best place to apply this offset is usually in the flash chip driver itself, as shown in the hypothetical `write_flash()` function in Listing 21-3.

```
#define PHYS_BASE_ADDRESS 0 // physical base address
#define ALIAS_BASE_ADDRESS 0x80000 // alias base address
void write_flash (long addr, unsigned char* data, int len)
{
    addr = (addr - PHYS_BASE_ADDRESS + ALIAS_BASE_ADDRESS);
    while (length) {
        ...
    }
}
```

Listing 21-3: Implementing memory aliasing in a write_flash() function.

Debugger Doesn't Like Self-relocating Code

One variation on the microprogrammer approach is to build the microprogrammer's functionality into the target system—a so-called *integral programmer*—instead of downloading it to RAM as the first step of the firmware update process. This strategy has its advantages, but it creates the need to copy the programmer's code from flash to RAM before the flash chip is erased and reprogrammed. In other words, the code must *self-relocate* to RAM at some point during its operation, so that it can continue to run after the flash chip is erased. This approach also requires that the code involved be position-independent.

The code in Listing 21-4 illustrates how to copy the integral programmer's code into RAM, and how to find the RAM version of the function `programmer()`. The symbols

RAM_PROG_START, PROG_LEN and ROM_PROG_START mark the regions in RAM and ROM where the programmer's code (which may be a subset of the total application) is located, and can often be computed automatically by the program's linker. The complicated-looking casting in the entrypoint calculation forces the compiler to do byte-sized address calculations when computing the value of entrypoint.

```
typedef int(*entrypoint_t)(void);
entrypoint_t relocate_programmer()
{
    entrypoint_t entrypoint;

    /* relocate the code */
    memcpy(RAM_PROG_START, ROM_PROG_START, PROG_LEN);

    /* find programmer() in ram: its location is the same
        offset from RAM_PROG_START as the rom version is
        from ROM_PROG_START */
    entrypoint = (entrypoint_t)((char*)programmer
        - (char*)ROM_PROG_START + (char*)RAM_PROG_START);

    return entrypoint;
}
```

Listing 21-4: Copying code into RAM.

When the caller invokes the function at the address returned by relocate_programmer(), control passes to the RAM copy of the microprogrammer code— and your debugger, if in use, stops showing any symbolic information related to the programmer() function. Why? Because programmer() is now running from an address that is different from the address it was originally located at, so the symbol information provided to the debugger by the linker is now meaningless.

One solution to this problem is to re-link the application with programmer() at the RAM address, and then import this symbol information into the debugger. This would be a convenient fix, except that not all debuggers support incremental additions to their symbol table. Another option is to simply suffer until the debugging of programmer() is complete, at which point you don't need to look at the code any more. In theory at least.

If the development environment is based on a hardware emulator rather than a self-hosted debugging agent, then you can completely avoid the hassles of code relocation by simply not relocating programmer() when an emulator is in use. When an emulator is present such relocation is, in fact, unnecessary: the opcodes associated with programmer() are actually located in the emulator's memory, rather than flash, so there is no worry that these instructions will disappear when the flash chip is erased and reprogrammed. This may also be the case for a self-hosted debugger setup, if the code being debugged and the debugging agent itself are both running from RAM.

Listing 21-5 illustrates an enhanced `relocate_programmer()` that does not copy code to RAM when an emulator is in use. Instead of using a #if compilation block to selectively enable or disable code copying, the function checks for an emulator at runtime, and skips the code relocation steps if it finds one.

```
typedef int (*entrypoint_t) (void);
entrypoint_t relocate_programmer()
{
    entrypoint_t entrypoint;

    /* test for an emulator, and only relocate code if necessary */
    if (memcmp(FLASH_START, FLASH_START + FLASH_SIZE, FLASH_SIZE))
        entrypoint = programmer;
    else {
        /* no emulator; copy programmer's memory section to ram */
        memcpy(RAM_PROG_START, ROM_PROG_START, PROG_LEN);
        entrypoint = (entrypoint_t) ((char*)programmer
            - (char*)ROM_PROG_START + (char*)RAM_PROG_START);
    }

    return entrypoint;
}
```

Listing 21-5: A smarter code relocation strategy,
which does not move code except when necessary.

The test for the presence of an emulator exploits the nature of the *memory alias* strategy discussed in the previous section. Without an emulator attached, the contents of memory in the region `FLASH_START` to `(FLASH_START+FLASH_SIZE)` must be identical to the memory in the region's memory alias, in this case assumed to start at `(FLASH_START+FLASH_SIZE)`, because the two address regions actually resolve to the same physical addresses in the flash chip.

With an emulator attached, however, a portion of flash memory is remapped to the emulator's internal memory space, so differences in comparisons can and do occur. These differences cause the `memcmp()` call to return nonzero, disclosing the presence of the emulator to `relocate_programmer()`. To improve performance, the region used for comparision can be reduced to just a handful of bytes that are known to change during compilation (a text string containing the time and date of the compilation, for example), or a known blank location in the flash chip that is preinitialized in the emulator's memory space to a value other than 0xff (the value of an erased flash memory cell).

Testing a blank flash memory cell also discloses the presence of an emulator when a memory alias of the flash chip is not available, which is useful in cases where the flash chip is so large that it occupies more than half the target processor's total address space— thereby making a complete memory alias impossible.

Can't generate position-independent code

Not all embedded development toolchains can produce code that is relocatable to arbitrary locations at runtime. Such position-*dependent* code must run from the address it was placed at by the linker, or the program will crash.

When a position-dependent microprogrammer is all that is available, then it obviously must be downloaded to the location in RAM that the linker intended it to run from. This implies that the memory reserved to hold the microprogrammer must be allocated at a known location.

With an integral programmer approach, there are two options. The first option is to compile the programmer code as a standalone program, located at its destination address in RAM. This code image is then included in the application image (perhaps by translating its binary image into a constant character array), and copied to RAM at the start of the firmware update process. This is a lot like a microprogrammer implementation, with the microprogrammer "downloaded" from the target's onboard memory instead of from a serial port.

The second option is to handle the integral programmer code as initialized data, and use the runtime environment's normal initialization procedures to copy the code into RAM. The GNU compiler supports this using its `__attribute__` language extension, and several commercial compilers provide this capability as well. The only limitation of this strategy is that it requires enough RAM space to hold the integral programmer code *plus* the program's other data.

No Firmware at Boot Time

Even in the most carefully designed downloadable firmware feature, the possibility still exists that the target hardware could attempt a startup using an accidentally blank flash chip. The outcome to avoid in this case is the sudden activation of the application's rotating machinery, which subsequently gobbles up an unprepared user.

The prescription for this case—which is best applied before a user is actually eaten—involves a careful study of the target processor's reaction to the illegal instructions and/or data represented by the `0xff`'s of an unprogrammed section of flash memory. Many embedded processors eventually halt processing, and tristate their control signals to let them float to whatever value is dictated by external hardware. Without pullup resistors or other precautions in place to force these uncontrolled signals to safe states, unpredictable and potentially lethal results are likely.

Persistent Watchdog Timeout

In systems that support downloadable firmware, an unavoidable application defect that forces a watchdog timeout and system reset can lock the microprogrammer out of the embedded system. The extreme case is an accidental `while(1);` statement in a program's `main()` function: the loop-within-a-loop that results (the infinite program loop, wrapped by the infinite watchdog timeout and system reset loop) keeps the target from responding to the **UPGRADE** button, because the system restarts before the button is even checked.

Systems that support downloadable firmware must carefully examine all available status circuitry to determine the reason the system is starting up, and force a transition to **UP-GRADE** mode in the event that an excessive number of watchdog or other resets are detected. Many embedded systems do not interrupt power to RAM when a watchdog timeout occurs, so it is safe to store a "magic number" and count of the number of resets there; the counter is incremented on each reset, and once a certain number is reached the system halts the application in an effort to avoid the code that forces the reset.

Unexpected Power Interruption

If power is lost unexpectedly during flash reprogramming, then the target's flash chip is left in a potentially inconsistent state when power is restored: maybe the programming operation finished and everything is fine, but probably not. The best case is when the system's boot code is intact, but portions of application code are missing; the worst case is where the flash chip is completely blank.

In the first case, a checksum can be used to detect the problem, and the system can force a transition to **UPGRADE** mode whether the user is requesting one or not. The only solution to the second case is to avoid ever having it happen.

One way to avoid producing a completely blank flash chip is to never erase the section of flash that contains the system's boot and programmer firmware. By leaving this code intact at all times, a hopefully-adequate environment is maintained even if power is interrupted. This approach may not always be an option, however: the flash chip may be a single-sector part that can only do an all-or-nothing erase, or the "boot sector" of the flash chip may be larger than the boot code it contains, and the wasted space is needed for application code.

A careful strategy for erasing and reprogramming the boot region of a flash chip can minimize the risk of damage to a system from unexpected power interruption, and in some cases eliminate it entirely.

The strategy works as follows: when a request to reprogram the section of flash containing boot code is detected, the programmer first creates a copy of the position-independent code in that section section to another section in flash, or assumes that a pre-positioned clone of the system's boot code exists. The boot code is then erased, and the target's reset vector is quickly restored to point to the temporary copy of the boot code. Once the boot sector is programmed, the reset vector is rewritten to point to the new startup code.

One of the two keys to the success of this strategy hinges on careful selection of the addresses used for the temporary and permanent copies of the boot code. If the permanent address is lower than the temporary address by a power of two, then the switch from the temporary reset vector to the permanent one can take place by only changing bits from one to zero, which makes a new erase of the flash sector unnecessary. This makes the switch from the temporary reset vector to the permanent one an atomic operation: no matter when power is interrupted, the vector still points to either one location or the other.

Obviously, the other critical element of this strategy is to eliminate the risk of power interruption in the moment between when the boot sector is erased and when the temporary reset vector is written. The amount of energy required to complete this operation can be computed by looking at the power consumption and timing specifications in the datasheet for the flash chip and microprocessor, and is usually in the range where additional capacitance in the system's power supply circuitry can bridge the gap if a power loss occurs. By checking that power has not been already lost at the start of the operation, and running without interruption until the sector is erased and the temporary reset vector is written, the remaining opportunity for damage is eliminated.

The limitation of this strategy is that it depends on a microprocessor startup process that reads a reset vector to get the value of the initial program counter. In processors that simply start running code from a fixed address after a reset, it may be possible to modify this strategy to accomplish the same thing with clever combinations of jmp and similar opcodes.

Hardware Alternatives

The microprogrammer-based firmware update strategy and its variations are all firmware-based, because they require that some code exist in the target system before that code can be reprogrammed. This creates a sort of chicken-and-egg problem: if you need code in the embedded system to put new code into the embedded system, how do you get the initial code there in the first place?

At least two existing hardware-based methods can be used to jump-start this process: *BDM* and *JTAG*.

A processor with a BDM port provides what is in essence a serial port tied directly to the guts of the microprocessor itself. By sending the right commands and data through this port, you can push a copy of your microprogrammer into the target's RAM space and hand over control to it. The BDM port can also be used to stimulate the i/o lines of the flash chip, thus programming it directly. Many BDM-based development systems include scripts and programs that implement this functionality, but it can also be implemented by hand with a few chips, a PC's printer port, and a careful study of the processor's datasheets.

JTAG is a fundamentally different technology designed to facilitate reading and writing of a chip's I/O lines, usually while the chip's microprocessor (if it has one) is held in reset. Like BDM, however, this capability can be used to stimulate a RAM or flash chip to push a microprogrammer application into it. And also like BDM, a JTAG interface can be built with just a few components and some persistent detective work in the target processor's manual.

A JTAG bus transceiver chip, versions of which are available from several vendors, can be added to systems that lack JTAG support.

Separating Code and Data

The ultimate goal of any downloadable firmware implementation effort is exactly that: a working downloadable firmware feature. Once this capability is safely in place, however, it is important to consider some other capabilities that the system can suddenly offer.

By separating an application's code and data into separate flash sectors, the possibility exists to update the two independently. This is useful if an application uses custom data like tuned parameters or accumulated measurements that cannot be easily replaced if lost. Such data tables must contain version information, however, so that later versions of an application can read old table formats—and so that old applications are not confused by new table formats.

The code in Listing 21-6 demonstrates one way to define a data table containing version information, and how to select from one of several data table formats at runtime.

Flexible and Safe

A microprogrammer-based downloadable firmware feature, when properly implemented, can safely add considerable flexiblity to an embedded system that uses flash memory. The techniques described here will help you avoid common mistakes and reap the uncommon benefits that microprogrammers and flash memory can offer.

```
/* the original table format,
   a.k.a. version 1 */
typedef struct {
  int x;
  int y;
} S_data_ver_1;

/* version 2 of the table format,
   which adds a 'z' field */
typedef struct {
  int x;
  int y;
  int z;
} S_data_ver_2;

/* the data, which always starts
   with a version identifier */
typedef struct{
  const int ver;
  union {
    S_data_ver_1 olddata[N];
    S_data_ver_2 newdata[N];
  };
} data_table;

void foo ( data_table* dt )
{
  int x, y, z, wdata;
  S_data_ver_1* dv1;
  S_data_ver_2* dv2;

  switch(dt->ver) {
  case 1:
    for( wdata = 0,
         dv1 = dt->olddata;
         wdata < N; wdata++, dv1++ ) {
      x = dv1->x;
      y = dv1->y;
      /* old data format did not include 'z',
         impose a default value */
      z = 0;
    }
    break;
  case 2:
    for( wdata = 0,
         dv2 = dt->newdata;
         wdata < N; wdata++, dv2++ ) {
      x = dv2->x;
      y = dv2->y;
      z = dv2->z;
    }
    break;
  default:
    /* unsupported format,
       select reasonable defaults */
    x = y = z = 0;
  }
}
```

Listing 21-6: Supporting multiple data table formats

Memory Diagnostics

Jack Ganssle

In "A Day in the Life" John Lennon wrote, "He blew his mind out in a car; he didn't notice that the lights had changed." As a technologist this always struck me as a profound statement about the complexity of modern life. Survival in the big city simply doesn't permit even a very human bit of daydreaming. Twentieth century life means keeping a level of awareness and even paranoia that our ancestors would have found inconceivable.

Since this song's release in 1967, survival has become predicated on much more than the threat of a couple of tons of steel hurtling though a red light. Software has been implicated in many deaths—plane crashes, radiation overexposures, pacemaker misfires. Perhaps a single bit, something so ethereal that it is nothing more than the charge held in an impossibly small well, is incorrect—that's all it takes to crash a system. Today's version of the Beatles song might include the refrain "He didn't notice that the bit had flipped."

Beyond software errors lurks the specter of a hardware failure that causes our correct code to die. Many of us write diagnostic code to help contain the problem.

ROM Tests

It doesn't take much to make at least the kernel of an embedded system run. With a working CPU chip, memories that do their thing, perhaps a dash of decoder logic, you can count on the code starting off... perhaps not crashing until running into a problem with I/O.

Though the kernel may be relatively simple, with the exception of the system's power supply it's by far the most intolerant portion of an embedded system to any sort of failure. The tiniest glitch, a single bit failure in a huge memory array, or any problem with the processor pretty much guarantees that nothing in the system stands a change of running.

Non-kernel failures may not be so devastating. Some I/O troubles will cause just part of the system to degrade, leaving much of the rest up. My car's black box seems to have forgotten how to run the cruise control, yet it still keeps the fuel injection and other systems running.

In the minicomputer era, most booted with a CPU test that checked each instruction. That level of paranoia is no longer appropriate, as a highly integrated CPU will generally fail disastrously. If the processor can execute any sort of a self test, it's pretty much guaranteed to be intact.

Dead decoder logic is just as catastrophic. No code will execute if the ROMs can't be selected.

If your boot ROM is totally misprogrammed or otherwise non-functional, then there's no way a ROM test will do anything other than crash. The value of a ROM test is limited to dealing with partially programmed devices (due, perhaps, to incomplete erasure, or inadvertently removing the device before completion of programming).

There's a small chance that ROM tests will pick up an addressing problem, if you're lucky enough to have a failure that leaves the boot and ROM test working. The odds are against it, and somehow Mother Nature tends to be very perverse.

Some developers feel that a ROM checksum makes sense to insure the correct device is inserted. This works best only if the checksum is stored outside of the ROM under test. Otherwise, inserting a device with the wrong code version will not show an error, as presumably the code will match the (also obsolete) checksum.

In multiple-ROM systems a checksum test can indeed detect misprogrammed devices, assuming the test code lives in the boot ROM. If this one device functions, and you write the code so that it runs without relying on any other ROM, then the test will pick up many errors.

Checksums, though, are passé. It's pretty easy for a couple of errors to cancel each other out. Compute a CRC (Cyclic Redundancy Check), a polynomial with terms fed back at various stages. CRCs are notoriously misunderstood but are really quite easy to implement. The best reference I have seen to date is "A Painless Guide to CRC Error Detection Algorithms", by Ross Williams. It's available via anonymous FTP from ftp.adelaide.edu.au/pub/rocksoft/crc_v3.txt.

The following code computes the 16 bit CRC of a ROM area (pointed to by rom, of size length) using the $x^{16} + x^{12} + x^5 + 1$ CRC:

```
#define CRC_P 0x8408
WORD rom_crc(char *rom, WORD length)
{
    unsigned char i;
    unsigned int value;
    unsigned int crc = 0xffff;

    do
    {
        for (i=0, value=(unsigned int)0xff & *rom++;
            i < 8;
            i++, value >>= 1)
        {
            if ((crc & 0x0001) ^ (value & 0x0001))
                    crc = (crc >> 1) ^ CRC_P;
            else   crc >>= 1;
        }
    } while (−length);
```

```
        crc = ~crc;
        value = crc;
        crc = (crc << 8) | ((value >> 8) & 0xff);

        return (crc);
    }
```

It's not a bad idea to add death traps to your ROM. On a Z80 0xff is a call to location 38. Conveniently, unprogrammed areas of ROMs are usually just this value. Tell your linker to set all unused areas to 0xff; then, if an address problem shows up, the system will generate lots of spurious calls. Sure, it'll trash the stack, but since the system is seriously dead anyway, who cares? Technicians can see the characteristic double write from the call, and can infer pretty quickly that the ROM is not working.

Other CPUs have similar instructions. Browse the op code list with a creative mind.

RAM Tests

Developers often adhere to beliefs about the right way to test RAM that are as polarized as disparate feelings about politics and religion. I'm no exception, and happily have this forum for blasting my own thoughts far and wide... so will shamelessly do so.

Obviously, a RAM problem will destroy most embedded systems. Errors reading from the stack will sure crash the code. Problems, especially intermittent ones, in the data areas may manifest bugs in subtle ways. Often you'd rather have a system that just doesn't boot, rather than one that occasionally returns incorrect answers.

Some embedded systems are pretty tolerant of memory problems. We hear of NASA spacecraft from time to time whose core or RAM develops a few bad bits, yet somehow the engineers patch their code to operate around the faulty areas, uploading the corrections over the distances of billions of miles.

Most of us work on systems with far less human intervention. There are no teams of highly trained personnel anxiously monitoring the health of each part of our products. It's our responsibility to build a system that works properly when the hardware is functional.

In some applications, though, a certain amount of self-diagnosis either makes sense or is required; critical life support applications should use every diagnostic concept possible to avoid disaster due to a sub-micron RAM imperfection.

So, my first belief about diagnostics in general, and RAM tests in particular, is to define your goals clearly. Why run the test? What will the result be? Who will be the unlucky recipient of the bad news in the event an error is found, and what do you expect that person to do?

Will a RAM problem kill someone? If so, a very comprehensive test, run regularly, is mandatory.

Is such a failure merely a nuisance? For instance, if it keeps a cell phone from booting, if there's nothing the customer can do about the failure anyway, then perhaps there's no reason for doing a test. As a consumer I could care less why the damn phone stopped working… if it's dead I'll take it in for repair or replacement.

Is production test—or even engineering test—the real motivation for writing diagnostic code? If so, then define exactly what problems you're looking for and write code that will find those sorts of troubles.

Next, inject a dose of reality into your evaluation. Remember that today's hardware is often very highly integrated. In the case of a microcontroller with on-board RAM the chances of a memory failure that doesn't also kill the CPU is small. Again, if the system is a critical life support application it may indeed make sense to run a test as even a minuscule probability of a fault may spell disaster.

Does it make sense to ignore RAM failures? If your CPU has an illegal instruction trap, there's a pretty good chance that memory problems will cause a code crash you can capture and process. If the chip includes protection mechanisms (like the x86 protected mode), count on bad stack reads immediately causing protection faults your handlers can process. Perhaps RAM tests are simply not required given these extra resources.

Too many of us use the simplest of tests—writing alternating 0x55 and 0xAA values to the entire memory array, and then reading the data to ensure it remains accessible. It's a seductively easy approach that will find an occasional problem (like, someone forgot to load all of the RAM chips), but that detects few real world errors.

Remember that RAM is an array divided into columns and rows. Accesses require proper chip selects and addresses sent to the array—and not a lot more. The 0x55/0xAA symmetrical pattern repeats massively all over the array; accessing problems (often more common than defective bits in the chips themselves) will create references to incorrect locations, yet almost certainly will return what appears to be correct data.

Consider the physical implementation of memory in your embedded system. The processor drives address and data lines to RAM—in a 16 bit system there will surely be at least 32 of these. Any short or open on this huge bus will create bad RAM accesses. Problems with the PC board are far more common than internal chip defects, yet the 0x55/0xAA test is singularly poor at picking up these, the most likely failures.

Yet, the simplicity of this test and its very rapid execution have made it an old standby used much too often. Isn't there an equally simple approach that will pick up more problems?

If your goal is to detect the most common faults (PCB wiring errors and chip failures more substantial than a few bad bits here or there), then indeed there is. Create a short string of almost random bytes that you repeatedly send to the array until all of memory is written. Then, read the array and compare against the original string.

I use the phrase "almost random" facetiously, but in fact it little matters what the string is, as long as it contains a variety of values. It's best to include the pathological cases, like 00, 0xaa, 0x55, and 0xff. The string is something you pick when writing the code, so it is truly not random, but other than these four specific values you fill the rest of it with nearly any set of values, since we're just checking basic write/read functions (remember: memory tends to fail in fairly dramatic ways). I like to use very orthogonal values—those with lots of bits changing between successive string members—to create big noise spikes on the data lines.

To make sure this test picks up addressing problems, ensure the string's length is not a factor of the length of the memory array. In other words, you don't want the string to be aligned on the same low-order addresses, which might cause an address error to go undetected. Since the string is much shorter than the length of the RAM array, you ensure it repeats at a rate that is not related to the row/column configuration of the chips.

For 64k of RAM, a string 257 bytes long is perfect. 257 is prime, and its square is greater than the size of the RAM array. Each instance of the string will start on a different low order address. 257 has another special magic: you can include every byte value (00 to 0xff) in the string without effort. Instead of manually creating a string in your code, build it in real time by incrementing a counter that overflows at 8 bits.

Critical to this, and every other RAM test algorithm, is that you write the pattern to all of RAM before doing the read test. Some people like to do non-destructive RAM tests by testing one location at a time, then restoring that location's value, before moving on to the next one. Do this and you'll be unable to detect even the most trivial addressing problem.

This algorithm writes and reads every RAM location once, so is quite fast. Improve the speed even more by skipping bytes, perhaps writing and reading every 3rd or 5th entry. The test will be a bit less robust yet will still find most PCB and many RAM failures.

Some folks like to run a test that exercises each and every bit in their RAM array. Though I remain skeptical of the need since most semiconductor RAM problems are rather catastrophic, if you do feel compelled to run such a test, consider adding another iteration of the algorithm just described, with all of the data bits inverted.

Sometimes, though, you'll want a more thorough test, something that looks for difficult hardware problems at the expense of speed.

When I speak to groups I'll often ask "What makes you think the hardware *really* works?" The response is usually a shrug of the shoulders, or an off-the-cuff remark about everything seeming to function properly, more or less, most of the time.

These qualitative responses are simply not adequate for today's complex systems. All too often, a prototype that seems perfect harbors hidden design faults that may only surface after you've built a thousand production units. Recalling products due to design bugs is unfair to the customer and possibly a disaster to your company.

Assume the design is absolutely ridden with problems. Use reasonable methodologies to find the bugs before building the first prototype, but then use that first unit as a testbed to find the rest of the latent troubles.

Large arrays of RAM memory are a constant source of reliability problems. It's indeed quite difficult to design the perfect RAM system, especially with the minimal margins and high speeds of today's 16 and 32 bit systems. If your system uses more than a couple of RAM parts, count on spending some time qualifying its reliability via the normal hardware diagnostic procedures. Create software RAM tests that hammer the array mercilessly.

Probably one of the most common forms of reliability problems with RAM arrays is pattern sensitivity. Now, this is not the famous pattern problems of yore, where the chips (particularly DRAMs) were sensitive to the groupings of ones and zeroes. Today the chips are just about perfect in this regard. No, today pattern problems come from poor electrical characteristics of the PC board, decoupling problems, electrical noise, and inadequate drive electronics.

PC boards were once nothing more than wiring platforms, slabs of tracks that propagated signals with near perfect fidelity. With very high speed signals, and edge rates (the time it takes a signal to go from a zero to a one or back) under a nanosecond, the PCB itself assumes all of the characteristics of an electronic component—one whose virtues are almost all problematic. It's a big subject (refer to read "High Speed Digital Design—a Handbook of Black Magic" by Howard Johnson and Martin Graham (1993 PTR Prentice Hall, NJ) for the canonical words of wisdom on this subject), but suffice to say a poorly designed PCB will create RAM reliability problems.

Equally important are the decoupling capacitors chosen, as well as their placement. Inadequate decoupling will create reliability problems as well.

Modern DRAM arrays are massively capacitive. Each address line might drive dozens of chips, with 5 to 10 pf of loading per chip. At high speeds the drive electronics must somehow drag all of these pseudo-capacitors up and down with little signal degradation. Not an easy job! Again, poorly designed drivers will make your system unreliable.

Electrical noise is another reliability culprit, sometimes in unexpected ways. For instance, CPUs with multiplexed address/data buses use external address latches to demux the bus. A signal, usually named ALE (Address Latch Enable) or AS (Address Strobe) drives the clock to these latches. The tiniest, most miserable amount of noise on ALE/AS will surely, at the time of maximum inconvenience, latch the data part of the cycle instead of the address. Other signals are also vulnerable to small noise spikes.

Many run-of-the-mill RAM tests, run for several hours, as you cycle the product through it's design environment (temperature, etc) will show intermittent RAM problems. These are symptoms of the design faults I've described, and always show a need for more work on the product's engineering.

Unhappily, all too often the RAM tests show no problem when hidden demons are indeed lurking. The algorithm I've described, as well as most of the others commonly used, tradeoff speed versus comprehensiveness. They don't pound on the hardware in a way designed to find noise and timing problems.

Digital systems are most susceptible to noise when large numbers of bits change all at once. This fact was exploited for data communications long ago with the invention of the Gray

Code, a variant of binary counting, where no more than one bit changes between codes. Your worst nightmares of RAM reliability occur when all of the address and/or data bits change suddenly from zeroes to ones.

For the sake of engineering testing, write RAM test code that exploits this known vulnerability. Write 0xffff to 0x0000 and then to 0xffff, and do a read-back test. Then write zeroes. Repeat as fast as your loop will let you go.

Depending on your CPU, the worst locations might be at 0x00ff and 0x0100, especially on 8 bit processors that multiplex just the lower 8 address lines. Hit these combinations, hard, as well.

Other addresses often exhibit similar pathological behavior. Try 0x5555 and 0xaaaa, which also have complementary bit patterns.

The trick is to write these patterns back-to-back. Don't test all of RAM, with the understanding that both 0x0000 and 0xffff will show up in the test. You'll stress the system most effectively by driving the bus massively up and down all at once.

Don't even think about writing this sort of code in C. Any high level language will inject too many instructions between those that move the bits up and down. Even in assembly the processor will have to do fetch cycles from wherever the code happens to be, which will slow down the pounding and make it a bit less effective.

There are some tricks, though. On a CPU with a prefetcher (all x86, 68k, etc.) try to fill the execution pipeline with code, so the processor does back-to-back writes or reads at the addresses you're trying to hit. And, use memory-to-memory transfers when possible. For example:

```
mov  si,0xaaaa
mov  di,0x5555
mov  [si],0xff
mov  [di],[si]
```

Nonvolatile Memory

Jack Ganssle

Many of the embedded systems that run our lives try to remember a little bit about us, or about their application domain, despite cycling power, brownouts, and all of the other perils of fixed and mobile operation. In the bad old days before microprocessors we had core memory, a magnetic medium that preserved its data when powered or otherwise.

Today we face a wide range of choices. Sometimes Flash or EEPROM is the natural choice for nonvolatile applications. Always remember, though, that these devices have limited numbers of write cycles. Worse, in some cases writes can be very slow.

Battery-backed up RAMs still account for a large percentage of nonvolatile systems. With robust hardware and software support they'll satisfy the most demanding of reliability fanatics; a little less design care is sure to result in occasional lost data.

Supervisory Circuits

In the early embedded days we were mostly blissfully unaware of the perils of losing power. Virtually all reset circuits were nothing more than a resistor/capacitor time constant. As Vcc ramped from 0 to 5 volts, the time constant held the CPU's reset input low—or lowish—long enough for the system's power supply to stabilize at 5 volts.

Though an elegantly simple design, RC time constants were flawed on the back end, when power goes away. Turn the wall switch off, and the 5 volt supply quickly decays to zero. Quickly only in human terms, of course, as many milliseconds went by while the CPU was powered by something between 0 and 5. The RC circuit is, of course, at this point at a logic one (not-reset), so allows the processor to run.

And run they do! With Vcc down to 3 or 4 volts most processors execute instructions like mad. Just not the ones you'd like to see. Run a CPU with out-of-spec power and expect random operation. There's a good chance the machine is going wild, maybe pushing and calling and writing and generally destroying the contents of your battery backed up RAM.

Worse, brown-outs, the plague of summer air conditioning, often cause small dips in voltage. If the AC mains decline to 80 volts for a few seconds a power supply might still crank out a few volts. When AC returns to full rated values the CPU is still running, back at 5 volts, but now horribly confused. The RC circuit never notices the dip from 5 to 3 or so volts, so the poor CPU continues running in its mentally unbalanced state. Again, your RAM is at risk.

Motorola, Maxim, and others developed many ICs designed specifically to combat these problems. Though features and specs vary, these supervisory circuits typically manage the processor's reset line, battery power to the RAM, and the RAM's chip selects.

Given that no processor will run reliably outside of its rated Vcc range, the first function of these chips is to assert reset whenever Vcc falls below about 4.7 volts (on 5 volt logic). Unlike an RC circuit which limply drools down as power fails, supervisory devices provide a snappy switch between a logic zero and one, bringing the processor to a sure, safe stopped condition.

They also manage the RAM's power, a tricky problem since it's provided from the system's Vcc when power is available, and from a small battery during quiescent periods. The switchover is instantaneous to keep data intact.

With RAM safely provided with backup power and the CPU driven into a reset state, a decent supervisory IC will also disable all chip selects to the RAM. The reason? At some point after Vcc collapses you can't even be sure the processor, and your decoding logic, will not create rogue RAM chip selects. Supervisory ICs are analog beasts, conceived outside of the domain of discrete ones and zeroes, and will maintain safe reset and chip select outputs even when Vcc is gone.

But check the specs on the IC. Some disable chip selects at exactly the same time they assert reset, asynchronously to what the processor is actually doing. If the processor initiates a write to RAM, and a nanosecond later the supervisory chip asserts reset and disables chip select, that write cycle will be one nanosecond long. *You cannot play with write timing and expect predictable results.* Allow any write in progress to complete before doing something as catastrophic as a reset.

Some of these chips also assert an NMI output when power starts going down. Use this to invoke your "oh_my_god_we're_dying" routine.

(Since processors usually offer but a single NMI input, when using a supervisory circuit never have any other NMI source. You'll need to combine the two signals somehow; doing so with logic is a disaster, since the gates will surely go brain dead due to Vcc starvation).

Check the specs on the parts, though, to ensure that NMI occurs *before* the reset clamp fires. Give the processor a handful of microseconds to respond to the interrupt before it enters the idle state.

There's a subtle reason why it makes sense to have an NMI power-loss handler: you want to get the CPU away from RAM. Stop it from doing RAM writes *before* reset occurs. If reset happens in the middle of a write cycle, there's no telling what will happen to your carefully protected RAM array. Hitting NMI first causes the CPU to take an interrupt exception, first finishing the current write cycle if any. This also, of course, eliminates troubles caused by chip selects that disappear synchronously to reset.

Every battery-backed up system should use a decent supervisory circuit; you just cannot expect reliable data retention otherwise. Yet, these parts are no panacea. The firmware itself is almost certainly doing things destined to defeat any bit of external logic.

Multi-byte Writes

There's another subtle failure mode that afflicts all too many battery-backed up systems. He observed that in a kinder, gentler world than the one we inhabit all memory transactions would require exactly one machine cycle, but here on Earth 8 and 16 bit machines constantly manipulate large data items. Floating point variables are typically 32 bits, so any store operation requires two or four distinct memory writes. Ditto for long integers.

The use of high-level languages accentuates the size of memory stores. Setting a character array, or defining a big structure, means that the simple act of assignment might require tens or hundreds of writes.

Consider the simple statement:

```
a=0x12345678;
```

An x86 compiler will typically generate code like:

```
mov [bx],5678
mov [bx+2],1234
```

which is perfectly reasonable and seemingly robust.

In a system with a heavy interrupt burden it's likely that sooner or later an interrupt will switch CPU contexts between the two instructions, leaving the variable "a" half-changed, in what is possibly an illegal state. This serious problem is easily defeated by avoiding global variables—as long as "a" is a local, no other task will ever try to use it in the half-changed state.

Power-down concerns twist the problem in a more intractable manner. As Vcc dies off a seemingly well-designed system will generate NMI while the processor can still think clearly. If that interrupt occurs during one of these multi-byte writes—as it eventually surely will, given the perversity of nature—your device will enter the power-shutdown code with data now corrupt. It's quite likely (especially if the data is transferred via CPU registers to RAM) that there's no reasonable way to reconstruct the lost data.

The simple expedient of eliminating global variables has no benefit to the power-down scenario.

Can you imagine the difficulty of *finding* a problem of this nature? One that occurs maybe once every several thousand power cycles, or less? In many systems it may be entirely reasonable to conclude that the frequency of failure is so low the problem might be safely ignored. This assumes you're not working on a safety-critical device, or one with mandated minimal MTBF numbers.

Before succumbing to the temptation to let things slide, though, consider implications of such a failure. Surely once in a while a critical data item will go bonkers. Does this mean your instrument might then exhibit an accuracy problem (for example, when the numbers are calibration coefficients)? Is there any chance things might go to an unsafe state? Does the loss of a critical communication parameter mean the device is dead until the user takes some presumably drastic action?

If the only downside is that the user's TV set occasionally—and rarely—forgets the last channel selected, perhaps there's no reason to worry much about losing multi-byte data. Other systems are not so forgiving.

Steve suggested implementing a data integrity check on power-up, to insure that no partial writes left big structures partially changed. I see two different directions this approach might take.

The first is a simple power-up check of RAM to make sure all data is intact. Every time a truly critical bit of data changes, update the CRC, so the boot-up check can see if data is intact. If not, at least let the user know that the unit is sick, data was lost, and some action might be required.

A second, and more robust, approach is to complete every data item write with a checksum or CRC of just that variable. Power-up checks of each item's CRC then reveals which variable was destroyed. Recovery software might, depending on the application, be able to fix the data, or at least force it to a reasonable value while warning the user that, whilst all is not well, the system has indeed made a recovery.

Though CRCs are an intriguing and seductive solution I'm not so sanguine about their usefulness. Philosophically it *is* important to warn the user rather than to crash or use bad data. But it's much better to never crash at all.

We can learn from the OOP community and change the way we write data to RAM (or, at least the critical items for which battery back-up is so important).

First, hide critical data items behind drivers. The best part of the OOP triptych mantra "encapsulation, inheritance, polymorphism" is "encapsulation." Bind the data items with the code that uses them. Avoid globals; change data by invoking a routine, a method, that does the actual work. Debugging the code becomes much easier, and reentrancy problems diminish.

Second, add a "`flush_writes`" routine to every device driver that handles a critical variable. "`Flush_writes`" finishes any interrupted write transaction. `Flush_writes` relies on the fact that only one routine—the driver—ever sets the variable.

Next, enhance the NMI power-down code to invoke all of the `flush_write` routines. Part of the power-down sequence then finishes all pending transactions, so the system's state will be intact when power comes back.

The downside to this approach is that you'll need a reasonable amount of time between detecting that power is going away, and when Vcc is no longer stable enough to support reliable processor operation. Depending on the number of variables needed flushing this might mean hundreds of microseconds.

Firmware people are often treated as the scum of the earth, as they inevitably get the hardware (late) and are still required to get the product to market on time. Worse, too many hardware groups don't listen to, or even solicit, requirements from the coding folks before cranking out PCBs. This, though, is a case where the firmware requirements clearly drive the hardware design. If the two groups don't speak, problems will result.

Some supervisory chips do provide advanced warning of imminent power-down. Maxim's (www.maxim-ic.com) MAX691, for example, detects Vcc failing below some value before shutting down RAM chip selects and slamming the system into a reset state. It also includes a separate voltage threshold detector designed to drive the CPU's NMI input when Vcc falls below some value you select (typically by selecting resistors). It's important to set this threshold above the point where the part goes into reset. Just as critical is understanding how power fails in your system. The capacitors, inductors, and other power supply components determine how much "alive" time your NMI routine will have before reset occurs. Make sure it's enough.

I mentioned the problem of power failure corrupting variables to Scott Rosenthal, one of the smartest embedded guys I know. His casual "yeah, sure, I see that all the time" got me interested. It seems that one of his projects, an FDA-approved medical device, uses hundreds of calibration variables stored in RAM. Losing any one means the instrument has to go back for readjustment. Power problems are just not acceptable.

His solution is a hybrid between the two approaches just described. The firmware maintains two separate RAM areas, with critical variables duplicated in each. Each variable has its own driver.

When it's time to change a variable, the driver sets a bit that indicates "change in process." It's updated, and a CRC is computed for that data item and stored with the item. The driver un-asserts the bit, and then performs the exact same function on the variable stored in the duplicate RAM area.

On power-up the code checks to insure that the CRCs are intact. If not, that indicates the variable was in the process of being changed, and is not correct, so data from the mirrored address is used. If both CRCs are OK, but the "being changed" bit is asserted, then the data protected by that bit is invalid, and correct information is extracted from the mirror site.

The result? With thousands of instruments in the field, over many years, not one has ever lost RAM.

Testing

Good hardware and firmware design leads to reliable systems. You won't know for sure, though, if your device really meets design goals without an extensive test program. Modern embedded systems are just too complex, with too much hard-to-model hardware/firmware interaction, to expect reliability without realistic testing.

This means you've got to pound on the product, and look for every possible failure mode. If you've written code to preserve variables around brown-outs and loss of Vcc, and don't conduct a meaningful test of that code, you'll probably ship a subtly broken product.

In the past I've hired teenagers to mindlessly and endlessly flip the power switch on and off, logging the number of cycles and the number of times the system properly comes to life. Though I do believe in bringing youngsters into the engineering labs to expose them to the cool parts of our profession, sentencing them to mindless work is a sure way to convince them to become lawyers rather than techies.

Better, automate the tests. The Poc-It, from Microtools (www.microtoolsinc.com/products.htm) is an indispensable $250 device for testing power-fail circuits and code. It's also a pretty fine way to find uninitialized variables, as well as isolating those awfully hard to initialize hardware devices like some FPGAs.

The Poc-It brainlessly turns your system on and off, counting the number of cycles. Another counter logs the number of times a logic signal asserts after power comes on. So, add a bit of test code to your firmware to drive a bit up when (and if) the system properly comes to life. Set the Poc-It up to run for a day or a month; come back and see if the number of power cycles is exactly equal to the number of successful assertions of the logic bit. Anything other than equality means something is dreadfully wrong.

Conclusion

When embedded processing was relatively rare, the occasional weird failure meant little. Hit the reset button and start over. That's less of a viable option now. We're surrounded by hundreds of CPUs, each doing its thing, each affecting our lives in different ways. Reliability will probably be the watchword of the next decade as our customers refuse to put up with the quirks that are all too common now.

The current drive is to add the maximum number of features possible to each product. I see cell phones that include games. Features are swell… if they work, if the product always fulfills its intended use. Cheat the customer out of reliability and your company is going to lose. Power cycling is something every product does, and is too important to ignore.

Proactive Debugging

Jack Ganssle

Academics who study software engineering have accumulated an impressive number of statistics about where bugs come from and how many a typical program will have once coding stops but before debugging starts. Amazingly, debugging burns about half the total engineering time for most products. That suggests minimizing debugging is the first place to focus our schedule optimization efforts.

Defect reduction is not rocket science. Start with a well-defined spec/requirements document, use a formal process of inspections on all work products (from the spec to the code itself), and give the developers powerful and reliable tools. Skip any of these and bugs will consume too much time and sap our spirits.

But no process, no matter how well defined, will eliminate all defects. We make mistakes!

Typically 5–10% of the source code will be wrong. Inspections catch 70% to 80% of those errors. A little 10,000 line program will still, after employing the very best software engineering practices, contain hundreds of bugs we've got to chase down before shipping. Use poor practices and the number (and development time) skyrockets.

Debugging is hard, slow, and frustrating. Since we know we'll have bugs, and we know some will be godawful hard to find, let's look for things we can do to our code to catch problems automatically or more easily. I call this *proactive debugging*, which is the art of anticipating problems and instrumenting the code accordingly.

Stacks and Heaps

Do you use the standard, well-known method for computing stack size? After all, undersize the stack and your program will crash in a terribly-hard-to-find manner. Allocate too much and you're throwing money away. High-speed RAM is expensive.

The standard stack-sizing methodology is to take a wild guess and hope. There is no scientific approach, nor even a rule of thumb. This isn't too awful in a single-task environment, since we can just stick the stack at the end of RAM and let it grow downwards, all the time hoping, of course, that it doesn't bang into memory already used. Toss in an RTOS, though, and such casual and sanguine allocation fails, since every task needs its own stack, each of which we'll allocate using the "take a guess and hope" methodology.

Take a wild guess and hope. Clearly this means we'll be wrong from time to time; perhaps even usually wrong. *If we're doing something that will likely be wrong, proactively take some action to catch the likely bug.*

Some RTOSes, for instance, include monitors that take an exception when any individual stack grows dangerously small. Though there's a performance penalty, consider turning these on for initial debug.

In a single task system, when debugging with an emulator or other tool with lots of hardware breakpoints, configure the tool to automatically (every time you fire it up) set a memory-write breakpoint near the end of the stack.

As soon as I allocate a stack I habitually fill each one with a pattern, like 0x55aa, even when using more sophisticated stack monitoring tools. After running the system for a minute or a week—whatever's representative for that application—I'll stop execution with the debugger and examine the stack(s). Thousands of words loaded with 0x55aa means I'm wasting RAM... and few or none of these words means the end is near, the stack's too small, and a crash is imminent.

Heaps are even more problematic. `Malloc()` is a nightmare for embedded systems. As with stacks, figuring the heap size is tough at best, a problem massively exacerbated by multitasking. `Malloc()` leads to heap fragmentation—though it may contain vast amounts of free memory, the heap may be so broken into small, unusable chunks that `malloc()` fails.

In simpler systems it's probably wise to avoid `malloc()` altogether. When there's enough RAM, allocating all variables and structures statically yields the fastest and most deterministic behavior, though at the cost of using more memory.

When dynamic allocation is unavoidable, by all means remember that `malloc()` has a return value! I look at a tremendous amount of firmware yet rarely see this function tested. It must be a guy thing. Testosterone. We're gonna `malloc` that puppy, by gawd, and that's that! Fact is, it may fail, which will cause our program to crash horribly. If we're smart enough—proactive enough—to test every `malloc()` then an allocation error will still cause the program to crash horribly, but at least we can set a debug trap, greatly simplifying the task of finding the problem.

An interesting alternative to `malloc()` is to use multiple heaps. Perhaps a heap for 100 byte allocations, one for 1000 bytes and another for 5000. Code a replacement `malloc()` that takes the heap identifier as its argument. Need 643 bytes? Allocate a 1000 byte block from the 1000 byte heap. Memory fragmentation becomes extinct, your code runs faster, though some RAM will be wasted for the duration of the allocation. A few commercial RTOSes do provide this sort of replacement `malloc()`.

Finally, if you do decide to use the conventional `malloc()`, at least for debugging purposes link in code to check the success of each allocation.

www.snippets.org/MEM.TXT (the link is case-sensitive) and companion files is Walter Bright's memory allocation test code, put into the public domain many years ago. MEM is a few hundred lines of C that replaces the library's standard memory functions with versions that diagnose common problems.

MEM looks for out-of-memory conditions, so if you've inherited a lot of poorly written code that doesn't properly check `malloc()`'s return value, use MEM to pick up errors. It verifies that frees match allocations. Before returning a block it sets the memory to a non-zero state to increase the likelihood that code expecting an initialized data set fails.

An interesting feature is that it detects pointer over- and under-runs. By allocating a bit more memory than you ask for, and writing a signature pattern into your pre- and post-buffer memory, when the buffer is freed MEM can check to see if a pointer wandered beyond the buffer's limits.

Geodesic Systems (www.geodesic.com) builds a commercial and much more sophisticated memory allocation monitor targeted at desktop systems. They claim that 99% of all PC programs suffer from memory leaks (mostly due to memory that is allocated but never freed). I have no idea how true this statement really is, but the performance of my PC sure seems to support their proposition. On a PC a memory leak isn't a huge problem, since the programs are either closed regularly or crash sufficiently often to let the OS reclaim that leaked resource.

Firmware, though, must run for weeks, months, even years without crashing. If 99% of PC apps suffer from leaks, I'd imagine a large number of embedded projects share similar problems. One of MEM's critical features is that it finds these leaks, generally before the system crashes and burns.

MEM is a freebie and requires only a small amount of extra code space, yet will find many classes of very common problems. The wise developer will link it, or other similar tools, into every project proactively, before the problems surface.

Seeding Memory

Bugs lead to program crashes. A "crash," though, can be awfully hard to find. First we notice a symptom—the system stops responding. If the debugger is connected, stopping execution shows that the program has run amok. But why? Hundreds of millions of instructions might elapse between ours seeing the problem and starting troubleshooting. No trace buffer is that large.

So we're forced to recreate the problem—if we can—and use various strategies to capture the instant when things fall apart. Yet "crash" often means that the code branches to an area of memory where it simply should not be. Sometimes this is within the body of code itself; often it's in an address range where there's neither code nor data.

Why do we continue to leave our unused ROM space initialized to some default value that's a function of the ROM technology and not what makes sense for us? Why don't we make a practice of setting all unused memory, both ROM and RAM, to a software interrupt instruction that immediately vectors execution to an exception handler?

Most CPUs have single byte or single word opcodes for a software interrupt. The Z80's RST7 was one of the most convenient, as it's 0xff which is the defaults state of unprogrammed EPROM. x86 processors all support the single byte INT3 software interrupt. Motorola's 68k family, and other processors, have an illegal instruction word.

Set all unused memory to the appropriate instruction, and write a handler that captures program flow if the software interrupt occurs. The stack often contains a wealth of clues about where things were and what was going on when the system crashed, so copy it to a debug area. In a multitasking application the OS's task control block and other data structures will have valuable hints. Preserve this critical tidbits of information.

Make sure the exception handler stops program flow; lock up in an infinite loop or something similar, ensure all interrupts and DMA are off, to stop the program from wandering away.

There's no guarantee that seeding memory will capture all crashes, but if it helps in even a third of the cases you've got a valuable bit of additional information to help diagnose problems.

But there's more to initializing memory than just seeding software interrupts. Other kinds of crashes require different proactive debug strategies. For example, a modern microprocessor might support literally hundreds of interrupt sources, with a vector table that dispatches ISRs for each. Yet the average embedded system might use a few, or perhaps a dozen, interrupts. What do we do with the unused vectors in the table?

Fill them, of course, with a vector aimed at an error handler! It's ridiculous to leave the unused vectors aimed at random memory locations. Sometime, for sure, you'll get a spurious interrupt, something awfully hard to track down. These come from a variety of sources, such as glitchy hardware (you're probably working on a barely functional hardware prototype, after all).

More likely is a mistake made in programming the vectors into the interrupting hardware. Peripherals have gotten so flexible that they're often impossible to manage. I've used parts with hundreds of internal registers, each of which has to be set just right to make the device function properly. Motorola's TPU, which is just a lousy timer, has a 142 page databook that documents some 36 internal registers. For a timer. I'm not smart enough to set them correctly first try, every time. Misprogramming any of these complex peripherals can easily lead to spurious interrupts.

The error handler can be nothing more than an infinite loop. Be sure to set up your debug tool so that every time you load the debugger it automatically sets a breakpoint on the handler. Again, this is nothing more than anticipating a tough problem, writing a tiny bit of code to capture the bug, and then configuring the tools to stop when and if it occurs.

Wandering Code

Embedded code written in any language seems determined to exit the required program flow and miraculously start running from data space or some other address range a very long way from code store. Sometimes keeping the code executing from ROM addresses feels like herding a flock of sheep, each of whom is determined to head off in its own direction.

In assembly a simple typo can lead to a jump to a data item; C, with support for function pointers, means state machines not perfectly coded might execute all over the CPU's address space. Hardware issues—like interrupt service routines with improperly initialized vectors and controllers—also lead to sudden and bizarre changes in program context.

Over the course of a few years I checked a couple of dozen embedded systems sent into my lab. The logic analyzer showed writes to ROM (surely an exercise in futility and a symptom of a bug) in more than half of the products.

Though there's no sharp distinction between wandering code and wandering pointers (as both often come from the same sorts of problems), diagnosing the problems requires different strategies and tools.

Quite a few companies sell products designed to find wandering code, or that can easily be adapted to this use. Some emulators, for instance, let you set up rules for the CPU's address space: a region might be enabled as execute-only, another for data read-writes but no executions, and a third tagged as no accesses allowed. When the code violates a rule the emulator stops, immediately signaling a serious bug. If your emulator includes this sort of feature, use it!

One of the most frustrating parts of being a tool vendor is that most developers use 10% of a tool's capability. We see engineers fighting difficult problems for hours, when a simple built-in feature might turn up the problem in seconds. I found that less than 1% of people I've worked with use these execution monitors, yet probably 100% run into crashes stemming from code flaws that the tools would pick up instantly.

Developers fall into four camps when using an execution monitoring device: the first bunch don't have the tool. Another group has one but never uses it, perhaps because they have simply not learned its fundamentals. To have unused debugging power seems a great pity to me. A third segment sets up and arms the monitoring tool only when it's obvious the code indeed wanders off somewhere, somehow.

The fourth, and sadly tiny, group builds a configuration file loaded by their ICE or debugger on every startup, that profiles what memory is where. These, in my mind, are the professional developers, the ones who prepare for disaster long before it inevitably strikes. Just like with make files, building configuration files takes tens of minutes so is too often neglected.

If your debugger or ICE doesn't come with this sort of feature, then adapt something else! A simple trick is to monitor the address bus with a logic analyzer programmed to look for illegal memory references. Set it to trigger on accesses to unused memory (most embedded systems use far less than the entire CPU address space; any access to an unused area indicates something is terribly wrong), or data-area executes, etc.

A couple of years ago I heard an interesting tale: an engineer, searching out errant writes, connected an analyzer to his system and quickly found the bug. Most of us would stop there, disconnect the tool, and continue traditional debugging. Instead, he left the analyzer connected for a number of weeks till completing debug. In that time he caught seven—count 'em—similar problems that may very well have gone undetected. These were weird writes to code or unused address spaces, bugs with no immediately apparent symptoms.

What brilliant engineering! He identified a problem, then developed a continuous process to always find new instances of the issue. Without this approach the unit would have shipped with these bugs undetected.

I believe that one of the most insidious problems facing firmware engineers are these lurking bugs. *Code that does things it shouldn't is flawed, even if the effects seem benign.*

Various studies suggest that *up to half the code in a typical system never gets tested.* Deeply nested ifs, odd exception/error conditions, and similar ills defy even well-designed test plans. A rule of thumb predicts that (uninspected) code has about a 5% error rate. This suggests that a 10,000 line project (not big by any measure) likely has 250 bugs poised to strike at the worst possible moment.

Special Decoders

Another option is to build a special PAL or PLD, connected to address and status lines, that flags errant bus transactions. Trigger an interrupt or a wait state when, say, the code writes to ROM.

If your system already uses a PAL, PLD or FPGA to decode memory chip selects, why not add an output that flags an error? The logic device—required anyway to enable memory banks—can also flash an LED or interrupt the CPU when it finds an error. A single added bit creates a complete self-monitoring system. More than just a great debug aid, it's a very intriguing adjunct to high-reliability systems. Cheap, too.

A very long time ago—way back in the 20th century, actually—virtually all microprocessor designs used an external programmable device to decode addresses. These designs were thus all ripe for this simple yet powerful addition. Things have changed; now CPUs sport sometimes dozens of chip select outputs. With so many selects the day of the external decoder is fading away.

Few modern design actually *use* all of those chip selects, though, opening up some interesting possibilities. Why not program one to detect accesses to unused memory? Fire off an interrupt or generate a wait state when such a condition occurs. The cost: zero. Effort needed: a matter of minutes. Considering the potential power of this oh-so-simple tool, the cost/benefit ratio is pretty stunning.

If you're willing to add a trivial amount of hardware, then exploit more of the unused chip selects. Have one monitor the bottom of the stack; AND it with `write` to detect stack overflows.

Be sure to AND the ROM chip select with `write` to find those silly code overwrites as well.

MMUs

If your system has an integrated memory management unit, more options exist. Many MMUs both translate addresses and provide memory protection features. Though MMUs are common only on higher end CPUs, where typically people develop pretty big applications, most of the systems I see pretty much bypass the MMU.

The classic example of this is protected mode in the x86 architecture. Intel's 16 bit x86 processors are all limited to a 1 Mb address range, using what's called "real mode." It's a way of getting direct access to 64k chunks of memory; with a little segment register, fiddling the entire 1Mb space is available.

But segmenting memory into 64k blocks gets awfully awkward as memory size increases. "Protected mode," Intel's answer to the problem, essentially lets you partition 4Gb of memory into thousands of variable size segments.

Protected mode resulted from a very reasonable design decision, that of preserving as much software compatibility as possible with the old real mode code. The ugly feelings about segments persisted, though, so even now most of the x86 protected mode embedded apps I see use map memory into a single, huge, 4 Gb segment. Doing this—a totally valid way of using the device's MMU—indeed gives the designer the nice flat address space we've all admired in the 68k family.

But it's a dumb idea.

The idea of segments, especially as implemented in protected mode, offers some stunning benefits. Break tasks into their own address spaces, with access rights for each. For the x86 MMU associates access rules for each segment. A cleverly written program can ensure that no one touches routines or data unless entitled to do so. The MMU checks each and every memory reference automatically, creating an easily-trapped exception when a violation occurs.

Most RTOSes targeted at big x86 applications explicitly support segmentation for just this reason. Me, I'd never build a 386 or bigger system without a commercial RTOS.

Beyond the benefits accrued from using protected mode to ensure data and code doesn't fall victim to cruddy code, why not use this very powerful feature for debugging?

First, do indeed plop code into segments whose attributes prohibit write accesses. Write an exception handler that traps these errors, and that logs the source of the error.

Second, map every unused area of memory into its own segment, which has access rights guaranteeing that any attempt to read or write from the region results in an exception.

It's a zero cost way to both increase system reliability in safety-critical applications, and to find those pesky bugs that might not manifest themselves as symptoms for a very long time.

Conclusion

With the exception of the logic analyzer—though a tool virtually all labs have anyway—all of the suggestions above require no tools. They are nearly zero-cost ways to detect a very common sort of problem. We know we'll have these sorts of problems; it's simply amateurish to not instrument your system *before* the problems manifest a symptom.

Solving problems is a high-visibility process; preventing problems is low-visibility. This is illustrated by an old parable:

In ancient China there was a family of healers, one of whom was known throughout the land and employed as a physician to a great lord. The physician was asked which of his family was the most skillful healer. He replied, "I tend to the sick and dying with drastic and dramatic treatments, and on occasion someone is cured and my name gets out among the lords."

"My elder brother cures sickness when it just begins to take root, and his skills are known among the local peasants and neighbors."

"My eldest brother is able to sense the spirit of sickness and eradicate it before it takes form. His name is unknown outside our home."

Great developers recognize that their code will be flawed, so instrument their code, and create toolchains designed to sniff out problems *before* a symptom even exists.

Exception Handling in C++
Here be Dragons
Alan Griffiths

Alan Griffiths is an experienced software developer with particular strengths in Object Oriented methods and technologies. These include Agile Methods, Java, C++, Object-Oriented Design and Generic Programming. He has a good working knowledge of tools and technologies like CVS, UML, HTML, XML, SQL, make, Ant, and JUnit. He has prepared and presented on technical subjects both internally to employers (Microlise and Experian) and at conferences, and has written feature articles for a range of technical publications.

"The use of animals in maps was commonplace from the earliest times. Man's deepest fears and superstitions concerning wide expanses of uncharted seas and vast tracts of 'terra incognita' were reflected in the monstrous animals that have appeared on maps ever since the Mappa Mundi." (Roderick Barron in "Decorative Maps")

For many developers C++ exception handling is like this—a Dark Continent with poor maps and rumors of ferocious beasts. I'm Alan Griffiths and I'm your guide to the landmarks and fauna of this region.

In order to discuss exception safety we need to cover a lot of territory. The next section identifies the "exception safe" mountains in the distance. Please don't skip it in the hope of getting to the good stuff—if you don't take the time to get your bearings now you'll end up in the wastelands.

Once I've established the bearings I'll show you a well-trodden path that leads straight towards the highest peak and straight through a tar pit. From experience, I've concluded that everyone has to go down this way once. So I'll go with you to make sure you come back. Not everyone comes back; some give up on the journey, others press on deeper and deeper into the tar pit until they sink from sight.

On our journey I'll tell you the history of how the experts sought for a long time before they discovered a route that bypasses that tar pit and other obstacles. Most maps don't show it yet, but I'll show you the signs to look out for. I'll also show you that the beasts are friendly and how to enlist their aid.

If you look into the distance you'll see a range of peaks, these are the heights of exception safety and are our final destination. But before we proceed on our trek let me point out two of the most important of these peaks, as we'll be using them as landmarks on our travels...

The Mountains (Landmarks of Exception Safety)

The difficulty in writing exception safe code isn't in writing the code that throws an exception, or in writing the code that catches the exception to handle it. There are many sources that cover these basics. I'm going to address the greater challenge of writing the code that lies in between the two.

Imagine for a moment the call stack of a running program: function a() has called function b(), b() has called c(), and so on, until we reach x(); x() encounters a problem and throws an exception. This exception causes the stack to unwind, deleting automatic variables along the way, until the exception is caught and dealt with by a().

I'm not going to spend any time on how to write functions a() or x(). I'm sure that the author of x() has a perfectly good reason for throwing an exception (running out of memory, disc storage, or whatever) and that the author of a() knows just what to do about it (display: "Sorry, please upgrade your computer and try again!").

The difficult problem is to write all the intervening functions in a way that ensures that something sensible happens as a result of this process. If we can achieve this we have "exception safe" code. Of course, that begs the question "what is something sensible?" To answer this let us consider a typical function f() in the middle of the call stack. How should f() behave?

Well, if f() were to handle the exception it might be reasonable for it to complete its task by another method (a different algorithm, or returning a "failed" status code). However, we are assuming the exception won't be handled until we reach a(). Since f() doesn't run to completion we might reasonably expect that:

1. f() doesn't complete its task.

2. If f() has opened a file, acquired a lock on a mutex, or, more generally; if f() has "allocated a resource" then the resource should not leak. (The file must be closed, the mutex must be unlocked, etc.)

3. If f() changes a data structure, then that structure should remain useable—e.g., no dangling pointers.

In summary: If f() updates the system state, then the state must remain valid. Note that isn't quite the same as correct—for example, part of an address may have changed leaving a valid address object containing an incorrect address.

I'm going to call these conditions the basic exception safety guarantee; this is the first, and smaller of our landmark mountains. Take a good look at it so that you'll recognize it later.

The basic exception safety guarantee may seem daunting but not only will we reach this in our travels, we will be reaching an even higher peak called the strong exception safety guarantee that places a more demanding constraint on f():

4. If f() terminates by propagating an exception then it has made no change to the state of the program.

Note that it is impossible to implement f() to deliver either the basic or strong exception safety guarantees if the behavior in the presence of exceptions of the functions it calls isn't known. This is particularly relevant when the client of f() (that is e()) supplies the functions to be called either as callbacks, as implementations of virtual member functions, or via template parameters. In such cases the only recourse is to document the constraints on them—as, for example, the standard library does for types supplied as template parameters to the containers.

If we assume a design with fully encapsulated data then each function need only be held directly responsible for aspects of the object of which it is a member. For the rest, the code in each function must rely on the functions it calls to behave as documented. (We have to rely on documentation in this case, since in C++ there is no way to express these constraints in the code.)

We'll rest here a while, and I'll tell you a little of the history of this landscape. Please take the time to make sure that you are familiar with these two exception safety guarantees. Later, when we have gained some altitude we will find that there is another peak in the mountain range: the no-throw exception safety guarantee—as the name suggests this implies that f() will never propagate an exception.

A History of this Territory

The C++ people first came to visit the land of exceptions around 1990 when Margaret Ellis and Bjarne Stroustrup published the "Annotated Reference Manual" [1]. Under the heading "experimental features" this described the basic mechanisms of exceptions in the language. In this early bestiary there is an early description of one of the friendly beasts we shall be meeting later on: it goes by the strange name of RESOURCE ACQUISITION IS INITIALIZATION.

By the time the ISO C++ Standards committee circulated "Committee Draft 1" in early 1995 C++ people were truly living in exception land. They hadn't really mapped the territory or produced an accurate bestiary but they were committed to staying and it was expected that these would soon be available.

However, by late 1996 when "Committee Draft 2" was circulated, the difficulties of this undertaking had become apparent. Around this time there came a number of reports from individual explorers. For example: Dave Abrahams identified the mountains we are using as landmarks in his paper "Exception Safety in STLPort" [2] although the basic exception safety guarantee was originally dubbed the "weak exception safety guarantee".

Some other studies of the region were produced by H Muller [3], Herb Sutter [4] and [5]. A little later came a sighting of another of the friendly beast that we will meet soon called ACQUISITION BEFORE RELEASE. This beast was first known by a subspecies named COPY BEFORE RELEASE and was identified by Kevlin Henney [6]. It is distinguished by the resources allocated being copies of dynamic objects.

By the time the ISO C++ Language Standard was published in 1998 the main tracks through the territory had been charted. In particular there are clauses in the standard guaranteeing the behavior of the standard library functions in the presence of exceptions. Also, in a number of key places within the standard, special mention is made of another friendly beast—SWAP in its incarnation as the `std::swap()` template function. We will be examining SWAP after our detour through the tar pit.

Since the publication of the ISO standard more modern charts through this landscape have been produced. Examples include the author's in an early version of this article [7]. A similar route is followed by Bjarne Stroustrup [8]. Herb Sutter [9] takes a different route, but the same landmarks are clearly seen.

OK, that's enough rest, we are going to take the obvious path and head directly towards the strong exception safety guarantee.

The Tar Pit

It is time to consider an example function, and for this part of the journey I have chosen the assignment operator for the following class:

```
class PartOne    { /* omitted */ };
class PartTwo    { /* omitted */ };

class Whole
{
public:
    // ...Lots omitted...

    Whole& operator=(const Whole& rhs);

private:
    PartOne*    p1;
    PartTwo*    p2;
};
```

Those of you that have lived in the old country will know the classical form for the assignment operator. It looks something like the following:

```
Whole& Whole::operator=(const Whole& rhs)
{
    if (&rhs != this)
    {
        delete p1;
        delete p2;
        p1 = new PartOne(*rhs.p1);
        p2 = new PartTwo(*rhs.p2);
    }
    return *this;
}
```

If you've not seen this before, don't worry because in the new land it is not safe. Either of the new expressions could reasonably throw (since at the very least they attempt to allocate memory) and this would leave the p1 and p2 pointers dangling. In theory the "delete" expressions could also throw—but in this article we will assume that destructors never propagate exceptions. (See: "destructors that throw exceptions.")

The obvious solution to the problems caused by an exception being propagated is to catch the exception and do some cleanup before throwing it again. After doing the obvious we have:

```cpp
Whole& Whole::operator=(const Whole& rhs)
{
    if (&rhs != this)
    {
        PartOne* t1 = new PartOne(*rhs.p1);

        try
        {
            PartTwo* t2 = new PartTwo(*rhs.p2);

            delete p1;
            delete p2;

            p1 = t1;
            p2 = t2;
        }
        catch (...)
        {
            delete t1;
            throw;
        }
    }
    return *this;
}
```

Let's examine why this works:

1. An exception in the first new expression isn't a problem—we haven't yet allocated any resources or modified anything.

2. If an exception is propagated from the second new expression, we need to release t1. So we catch it, delete t1 and throw the exception again to let it propagate.

3. We are assuming that destructors don't throw, so we pass over the two deletes without incident. Similarly the two assignments are of base types (pointers) and cannot throw an exception.

4. The state of the Whole isn't altered until we've done all the things that might throw an exception.

If you peer carefully through the undergrowth you can see the first of the friendly beasts. This one is called ACQUISITION BEFORE RELEASE. It is recognized because the code is organized so that new resources (the new PartOne and PartTwo) are successfully acquired before the old ones are released.

We've achieved the strong exception safety guarantee on our first attempt! But there is some black sticky stuff on our boots.

Tar!

There are problems lying just beneath the surface of this solution. I chose an example that would enable us to pass over the tar pit without sinking too deep. Despite this, we've incurred costs: the line count has doubled and it takes a lot more effort to understand the code well enough to decide that it works.

If you want to, you may take some time out to convince yourself of the existence of the tar pit—I'll wait. Try the analogous example with three pointers to parts or replacing the pointers with two parts whose assignment operators may throw exceptions. With real life examples, things get very messy very quickly.

Many people have reached this point and got discouraged. I agree with them: routinely writing code this way is not reasonable. Too much effort is expended on exception safety housekeeping chores like releasing resources. If you hear that "writing exception safe code is hard" or that "all those try...catch blocks take up too much space" you are listening to someone that has discovered the tar pit.

I'm now going to show you how exception handling allows you to use less code (not more), and I'm not going to use a single try...catch block for the rest of the article! (In a real program the exception must be caught somewhere—like function a() in the discussion above, but most functions simply need to let the exceptions pass through safely.)

The Royal Road

There are three "golden rules":

1. Destructors may not propagate exceptions,

2. The states of two instances of a class may be swapped without an exception being thrown,

3. An object may own at most one resource,

We've already met the first rule.

The second rule isn't obvious, but is the basis on which SWAP operates and is key to exception safety. The idea of SWAP is that for two instances of a class that owns resources exchanging the states is feasible without the need to allocate additional resources. Since nothing needs to be allocated, failure needn't be an option and consequently neither must throwing an exception. (It is worth mentioning that the no-throw guarantee is not feasible for assignment, which may have to allocate resources.)

If you look at the ISO C++ Language Standard, you'll find that `std::swap()` provides the no-throw guarantee for fundamental types and for relevant types in the standard library. This is achieved by overloading `std::swap()`—e.g., there is a template corresponding to each of the STL containers. This looks like a good way to approach SWAP but introducing additional overloads of `std::swap()` is not permitted by the language standard. The standard does permit to explicit specialization of an existing `std::swap()` template function on user defined classes and this is what I would recommend doing where applicable (there is an example below). The standards committee is currently considering a defect report that addresses the problem caused by these rules for the authors of user defined template classes. (See: Standard Algorithms and User Defined Template Classes.)

The third rule addresses the cause of all the messy exception handling code we saw in the last section. It was because creating a new second part might fail that we wrote code to handle it and doubled the number of lines in the assignment operator.

We'll now revisit the last example and make use of the above rules. In order to conform to the rule regarding ownership of multiple objects we'll delegate the responsibility of resource ownership to a couple of helper classes. I'm using the `std::auto_ptr<>` template to generate the helper classes here because it is standard, not because it is the ideal choice.

```
class Whole {
public:
    // ...Lots omitted...

    Whole& operator=(const Whole& rhs);

private:
    std::auto_ptr<PartOne>     p1;
    std::auto_ptr<PartTwo>     p2;
};


Whole& Whole::operator=(const Whole& rhs)
{
    std::auto_ptr<PartOne> t1(new PartOne(*rhs.p1));
    std::auto_ptr<PartTwo> t2(new PartTwo(*rhs.p2));

    std::swap(p1, t1);
    std::swap(p2, t2);

    return *this;
}
```

Not only is this shorter than the original exception-unsafe example, it meets the strong exception safety guarantee.

Look at why it works:

1. There are no leaks: whether the function exits normally, or via an exception, `t1` and `t2` will delete the parts they currently own.

2. The swap expressions cannot throw (second rule).

3. The state of the Whole isn't altered until we've done all the things that might throw an exception.

Oh, by the way, I've not forgotten about self-assignment. Think about it—you will see the code works without a test for self-assignment. Such a test may be a bad idea: assuming that self-assignment is very rare in real code and that the branch could have a significant cost, Francis Glassborow suggested a similar style of assignment operator as a speed optimization [10]. Following on from this, Kevlin Henney explored its exception safety aspects in [11], [12] and [6].

We are on much firmer ground than before: it isn't hard to see why the code works and generalizing it is simple. You should be able to see how to manage a Whole with three auto_ptrs to Parts without breaking stride.

You can also see another of the friendly beasts for the first time. Putting the allocation of a resource (here a new expression) into the initializer of an instance of a class (e.g., `auto_ptr<PartOne>`) that will delete it on destruction is RESOURCE ACQUISITION IS INITIALIZATION. And, of course, we can once again see ACQUISITION BEFORE RELEASE.

(Yes, in this case we could use assignment instead of SWAP to make the updates. However with a more complex type SWAP is necessary, as we shall see later. I use SWAP in this example for consistency.)

The Assignment Operator—A Special Case

Before I go on to deal with having members that may throw when updated, I've a confession I need to make. It is possible, and usual, to write the assignment operator more simply than the way I've just demonstrated. The above method is more general than what follows and can be applied when only some aspects of the state are being modified. The following applies only to assignment:

```
Whole& Whole::operator=(const Whole& rhs)
{
    Whole(rhs).swap(*this);
    return *this;
}
```

Remember the second rule: Whole is a good citizen and provides for SWAP (by supplying the `swap()` member function). I also make use of the copy constructor—but it would be a perverse class design that supported assignment but not copy construction. I'm not sure whether the zoologists have determined the relationship between SWAP and copying here, but the traveller won't go far wrong in considering COPY AND SWAP as a species in its own right.

For completeness, I'll show the methods used above:

```
void Whole::swap(Whole& that)
{
    std::swap(p1, that.p1);
    std::swap(p2, that.p2);
}

Whole::Whole(const Whole& rhs)
:   p1(new PartOne(*rhs.p1)),
    p2(new PartTwo(*rhs.p2))
{
}
```

One further point about making Whole a good citizen is that we need to specialize `std::swap()` to work through the `swap()` member function. By default `std::swap()` will use assignment—and not deliver the no-throw guarantee we need for SWAP. The standard allows us to specialize existing names in the std namespace on our own types, and it is good practice to do so in the header that defines the type.

```
namespace std
{
    template<>
    inline void swap(exe::Whole& lhs, exe::Whole& rhs)
    {
        lhs.swap(rhs);
    }
}
```

This avoids any unpleasant surprises for client code that attempts to `swap()` two Wholes.

Although we've focused on attaining the higher peak of strong exception safety guarantee, we've actually covered all the essential techniques for achieving either strong or basic exception safety. The remainder of the article shows the same techniques being employed in a more complex example and gives some indication of the reasons you might choose to approach the lesser altitudes of basic exception safety.

In Bad Weather

We can't always rely on bright sunshine, or on member variables that are as easy to manipulate as pointers. Sometimes we have to deal with rain and snow, or base classes and member variables with internal state.

To introduce a more complicated example, I'm going to elaborate the Whole class we've just developed by adding methods that update p1 and p2. Then I'll derive an ExtendedWhole class from it that also contains an instance of another class: PartThree. We'll be assuming that operations on PartThree are exception safe, but, for the purposes of discussion, I'll leave it open whether PartThree offers the basic or the strong exception safety guarantee.

```
Whole& Whole::setP1(const PartOne& value)
{
   p1.reset(new PartOne(value));
   return *this;
}


Whole& Whole::setP2(const PartTwo& value)
{
   p2.reset(new PartTwo(value));
   return *this;
}



class ExtendedWhole : private Whole
.{
public:

   // Omitted constructors & assignment

   void swap(const ExtendedWhole& rhs);

   void setParts(
      const PartOne& p1,
      const PartTwo& p2,
      const PartThree& p3);

private:
   int       count;
   PartThree body;
};
```

The examples we've looked at so far are a sufficient guide to writing the constructors and assignment operators. We are going to focus on two methods: the swap() member function and a setParts() method that updates the parts.

Writing swap() looks pretty easy—we just swap the base class, and each of the members. Since each of these operations is "no-throw" the combination of them is also "no-throw".

```
void ExtendedWhole::swap(ExtendedWhole& rhs)
{
   Whole::swap(rhs);
   std::swap(count, rhs.count);
   std::swap(body,  rhs.body);
}
```

Writing `setParts()` looks equally easy: Whole provides methods for setting `p1` and `p2`, and we have access to body to set that. Each of these operations is exception safe, indeed the only one that need not make the strong exception safety guarantee is the assignment to body. Think about it for a moment: is this version of `setParts()` exception safe? And does it matter if the assignment to body offers the basic or strong guarantee?

```
void ExtendedWhole::setParts(
    const PartOne& p1,
    const PartTwo& p2,
    const PartThree& p3)
{
    setP1(p1);
    setP2(p2);
    body = p3;
}
```

Let's go through it together. None of the operations leak resources, and `setParts()` doesn't allocate any so we don't have any leaks. If an exception propagates from any of the operations, then they leave the corresponding sub-object in a useable state, and presumably that leaves ExtendedWhole useable (it is possible, but in this context implausible, to construct examples where this isn't true). However, if an exception propagates from `setP2()` or from the assignment then the system state has been changed. And this is so regardless of which guarantee PartThree makes.

The simple way to support the strong exception safety guarantee is to ensure that nothing is updated until we've executed all the steps that might throw an exception. This means taking copies of sub-objects and making the changes on the copies, prior to swapping the state between the copies and the original sub-objects:

```
void ExtendedWhole::setParts(
    const PartOne& p1,
    const PartTwo& p2,
    const PartThree& p3)
{
    Whole temp(*this);
    temp.setP1(p1).setP2(p2);
    body = p3;
    Whole::swap(temp);
}
```

Once again does it matter if the assignment to body offers the basic or strong guarantee? Yes it does, if it offers the strong guarantee then all is well with the above, if not then the assignment needs to be replaced with COPY AND SWAP, viz:

```
PartThree(p3).swap(body);
```

Once again we have attained the highest peak, but this may not be healthy. On terrestrial mountains above a certain height there is a "death zone" where the supply of oxygen is insufficient to support life. Something similar happens with exception safety: there is a cost to implementing the strong exception safety guarantee. Although the code you write isn't much more complicated than the "basic" version, additional objects are created and these allocate resources at runtime. This causes the program to make more use of resources and to spend time allocating and releasing them.

Trying to remain forever at high altitude will drain the vitality. Fortunately, the basic exception safety guarantee is below the death zone: when one makes a composite operation whose parts offer this guarantee one automatically attains the basic guarantee (as the first version of `setParts()` shows this is not true of the strong guarantee). From the basic guarantee there is an easy climb from this level to the strong guarantee by means of COPY AND SWAP.

Looking back

Before we descend from the peak of strong exception safety guarantee and return to our starting point, look back over the route we covered. In the distance you can see the well-trampled path that led to the tar pit and just below us the few tracks leading from the tar pit up a treacherous slope to where we stand. Off to the left is the easier ridge path ascending from basic exception safety guarantee and beyond that the road that led us past the tar pit. Fix these landmarks in your mind and remember that the beasts we met are not as fierce as their reputations.

Destructors that throw exceptions.

Exceptions propagating from destructors cause a number of problems. For example, consider a Whole that holds pointers to a PartOne, a PartTwo, and a PartThree that it owns (ie it must delete them). If the destructors of the parts propagate exceptions, we would have trouble just writing a destructor for Whole. If more than one destructor throws, we must suppress at least one exception while remembering to destroy the third part. Writing update methods (like assignment) under such circumstances is prohibitively difficult or impossible.

There are many situations where an exception propagating from a destructor is extremely inconvenient—my advice is not to allow classes that behave in this manner into your system. (If forced to, you can always 'wrap' them in a well behaved class of your own.)

If you look at the standard library containers, you'll find that they place certain requirements on the types that are supplied as template parameters. One of these is that the destructor doesn't throw exceptions. There is a good reason for this: it is hard to write code that manipulates objects that throw exceptions when you try to destroy them. In many cases, it is impossible to write efficient code under such circumstances.

In addition, the C++ exception handling mechanism itself objects to destructors propagating exceptions during the "stack unwinding" process. Indeed, unless the application developer takes extraordinary precautions, the application will be terminated in a graceless manner.

There is no advantage in allowing destructors to propagate exceptions and a whole host of disadvantages. It should be easy to achieve: in most cases all a destructor should be doing is destroying other objects whose destructors shouldn't throw, or releasing resources—and if that fails an exception won't help.

Apart from the practicalities, what does an exception from a destructor mean? If I try to destroy an object and this fails what am I supposed to do? Try again?

Destructors that throw exceptions? Just say no.

Standard Algorithms and User Defined Template Classes.

The std::swap() template functions are one example of an algorithm implemented by the standard library. It is also an example where there is a good reason for C++ users to endeavour to provide an implementation specific to the needs of the classes and class templates that they develop. This need is particularly significant to developers of extension libraries—who would like to ensure that what they develop will both work well with the library and with other extension libraries. (*Continued on next page.*)

Standard Algorithms and User Defined Template Classes.

(*Continued from previous page.*) So consider the plight of a developer who is writing a template class Foo<> that takes a template parameter T and wishes to SWAP two instances of T. Now Foo is being instantiated with a fundamental type, or with an instance of any swappable type from the standard library the correct function can be resolved by writing:

```
Using std::swap;
Swap(t1, t2);
```

However, the primary template std::swap() that will be instantiated for other types is guaranteed to use copy construction and assignment unless an explicit specialization has been provided (and this is impractical if T is a specialization of a template class). As we have seen, copy construction and assignment probably won't meet the requirements of SWAP. Now this won't always matter, because a language mechanism "Argument-dependent name lookup" (commonly known as "Koenig Lookup") might introduce into overload resolution a function called swap() from the namespace in which T is declared. If this is the best match for arguments of type T then it is the one that gets called and the templates in std are ignored.

Now there are three problems with this:

1. Depending on the context of the above code, Koenig Lookup produces different results. (A library developer might reasonably be expected to know the implications of a class member named "swap" and how to deal with them—but many don't.) Most C++ users will simply get it wrong—without any obvious errors—when only the std::swap() templates are considered.

2. The standard places no requirements on functions called "swap" in any namespace other than std—so there is no guarantee that bar::swap() will do the right thing.

3. In a recent resolution of an issue the standards committee has indicated that where one standard function/template function is required to be used by another then the use should be fully qualified (i.e. std::swap(t1, t2);) to prevent the application of Koenig Lookup. If you (or I) provide yournamespace::swap() the standard algorithms won't use it.

Since the standards committee is still considering what to do about this I can't give you a watertight recommendation. I am hoping that in the short term a "technical corrigenda" will permit users to introduce new overloads of such template functions in std. So far as I know this technique works on all current implementations—if you know of one where it doesn't please let me know. In the longer term I am hoping that the core language will be extended to permit the partial specialization of template functions (and also that the library changes to use partial specialization in place of overloading).

The trouble with std::auto_ptr<>.

By historical accident, the standard library provides a single smart pointer template known as auto_ptr<>. auto_ptr<> has what I will politely describe as "interesting" copy semantics. Specifically, if one auto_ptr<> is assigned (or copy constructed) from another then they are both changed—the auto_ptr<> that originally owned the object loses ownership and becomes 0. This is a trap for the unwary traveller! There are situations that call for this behavior, but on most occasions that require a smart pointer the copy semantics cause a problem.

When we replace PartXXX* with auto_ptr<PartXXX> in the Whole class we still need to write the copy constructor and assignment operator carefully to avoid an PartXXX being passed from one Whole to another (with the consequence that one Whole loses its PartXXX).

Worse than this, if we attempt to hide the implementation of PartXXX from the client code using a forward declaration we also need to write the destructor. If we don't, the one the compiler generates for us will not correctly destroy the PartXXX. This is because the client code causes the generation of the Whole destructor and consequently instantiates the auto_ptr<> destructor without having seen the class definition for PartXXX. The effect of this is that the destructor of PartXXX is never called. (A good compiler may give a cryptic warning about deleting an incomplete type, but the language standard requires that the code compiles—although the results are unlikely to be what the programmer intended.)

Although the standard library doesn't support our needs very well it is possible to write smart pointers which do. There are a couple of examples in the "arglib" library on my website. (Both "arg::body_part_ptr<>" and "arg::grin_ptr<>" are more suitable than std::auto_ptr<>.)

The cost of exception handling.

Compiler support for exception handling does make the generated code bigger (figures vary around 10-15%), but only for the same code. However, code isn't written the same way without exceptions—for example, since constructors cannot return an error code, idioms such as "two phase construction" are required. I have here a comparable piece of code to the final example that has been handed down the generations from a time before the introduction of exception handling to C++. (Actually I've made it up—but I was around back then and remember working with code like this, so it is an authentic fake.)

```
int ExtendedWhole::setParts(
    const PartOne& p1,
    const PartTwo& p2,
    const PartThree& p3)
{

    Whole tw;
    int rcode = tw.init(*this);

    if (!rcode) rcode = tw.setP1(p1);

    if (!rcode) rcode = tw.setP2(p2);

    if (!rcode)
    {
        PartThree t3;
        rcode = t3.copy(p3);

        if (!rcode)
        {
            Whole::swap(tw);
            body.swap(t3);
        }
    }
    return rcode;
}
```

To modern eyes the need to repeat this testing & branch on return codes looks very like the tar-pit we encountered earlier—it is verbose, hard to validate code. I'm not aware of any trials where comparable code was developed using both techniques, but my expectation is that the saving in hand-written code from using exceptions significantly outweighs the extra cost in compiler-generated exception handling mechanisms.

Please don't take this as a rejection of return codes, they are one of the primary error reporting mechanisms in C++. But if an operation will only fail in exceptional circumstances (usually running out of a resource) or cannot reasonably be expected to be dealt with by the code at the call site then exceptions can greatly simplify the task.

References

1. The Annotated C++ Reference Manual
    Ellis & Stroustrup
    ISBN 0-201-51459-1

2. Exception Safety in STLPort
    Dave Abrahams
    http://www.stlport.org/doc/exception_safety.html

3. Ten Rules for Handling Exception Handling Successfully
    H. Muller
    C++ Report Jan.'96

4. Designing exception-safe Generic Containers
    Herb Sutter
    C++ Report Sept.'97

5. More Exception-safe Generic Containers
    Herb Sutter
    C++ Report Nov-Dec.'97

6. Creating Stable Assignments
    Kevlin Henney
    C++ Report June'98

7. The safe path to C++ exceptions
    Alan Griffiths
    EXE Dec.'99

8. The C++ Programming Language (3rd Edition) appendix E
    "Standard Library Exception Safety"
    Bjarne Stroustrup

9. Exceptional C++
    Herb Sutter
    ISBN 0-201-61562-2

10. The Problem of Self-Assignment
    Francis Glassborow   Overload 19
    ISSN 1354-3172

11. Self Assignment? No Problem!
    Kevlin Henney   Overload 20
    ISSN 1354-3172

12. Self Assignment? No Problem!
    Kevlin Henney   Overload 21
    ISSN 1354-3172

Building a Great Watchdog

Jack Ganssle

Launched in January of 1994, the Clementine spacecraft spent two very successful months mapping the moon before leaving lunar orbit to head towards near-Earth asteroid Geographos.

A dual-processor Honeywell 1750 system handled telemetry and various spacecraft functions. Though the 1750 could control Clementine's thrusters, it did so only in emergency situations; all routine thruster operations were under ground control.

On May 7 the 1750 experienced a floating point exception. This wasn't unusual; some 3000 prior exceptions had been detected and handled properly. But immediately after the May 7 event downlinked data started varying wildly and nonsensically. Then the data froze. Controllers spent 20 minutes trying to bring the system back to life by sending software resets to the 1750; all were ignored. A hardware reset command finally brought Clementine back on-line.

Alive, yes, even communicating with the ground, but with virtually no fuel left.

The evidence suggests that the 1750 locked up, probably due to a software crash. While hung the processor turned on one or more thrusters, dumping fuel and setting the spacecraft spinning at 80 RPM. In other words, it appears the code ran wild, firing thrusters it should never have enabled; they kept firing till the tanks ran nearly dry and the hardware reset closed the valves.

The mission to Geographos had to be abandoned.

Designers had worried about this sort of problem and implemented a software thruster timeout. That, of course, failed when the firmware hung.

The 1750's built-in watchdog timer hardware was not used, over the objections of the lead software designer. With no automatic "reset" button, success of the mission rested in the abilities of the controllers on Earth to detect problems quickly and send a hardware reset. For the lack of a few lines of watchdog code the mission was lost.

Though such a fuel dump had never occurred on Clementine before, roughly 16 times before the May 7 event hardware resets from the ground had been required to bring the spacecraft's firmware back to life. One might also wonder why some 3000 previous floating point exceptions were part of the mission's normal firmware profile.

Not surprisingly, the software team wished they had indeed used the watchdog, and had not implemented the thruster timeout in firmware. They also noted, though, that a normal, simple, watchdog may not have been robust enough to catch the failure mode.

Contrast this with Pathfinder, a mission whose software also famously hung, but which was saved by a reliable watchdog. The software team found and fixed the bug, uploading new code to a target system 40 million miles away, enabling an amazing roving scientific mission on Mars.

Watchdog timers (WDTs) are our failsafe, our last line of defense, an option taken only when all else fails—right? These missions (Clementine had been reset *16 times* prior to the failure) and so many others suggest to me that WDTs are not emergency outs, but integral parts of our systems. The WDT is as important as `main()` or the runtime library; it's an asset that is likely to be used, and maybe used a lot.

Outer space is a hostile environment, of course, with high intensity radiation fields, thermal extremes and vibrations we'd never see on Earth. Do we have these worries when designing Earth-bound systems?

Maybe so. Intel revealed that the McKinley processor's ultra fine design rules and huge transistor budget means cosmic rays may flip on-chip bits. The Itanium 2 processor, also sporting an astronomical transistor budget and small geometry, includes an on-board system management unit to handle transient hardware failures. The hardware ain't what it used to be—even if our software were perfect.

But too much (all?) firmware is not perfect. Consider this unfortunately true story from Ed VanderPloeg:

> "The world has reached a new embedded software milestone: I had to reboot my hood fan. That's right, the range exhaust fan in the kitchen. It's a simple model from a popular North American company. It has six buttons on the front: 3 for low, medium, and high fan speeds and 3 more for low, medium, and high light levels. Press a button once and the hood fan does what the button says. Press the same button again and the fan or lights turn off. That's it. Nothing fancy. And it needed rebooting via the breaker panel."

> "Apparently the thing has a micro to control the light levels and fan speeds, and it also has a temperature sensor to automatically switch the fan to high speed if the temperature exceeds some fixed threshold. Well, one day we were cooking dinner as usual, steaming a pot of potatoes, and suddenly the fan kicks into high speed and the lights start flashing. 'Hmm, flaky sensor or buggy sensor software,' I think to myself."

> "The food happened to be done so I turned off the stove and tried to turn off the fan, but I suppose it wanted things to cool off first. Fine. So after ten minutes or so the fan and lights turned off on their own. I then went to turn on the lights, but instead they flashed continuously, with the flash rate depending on the brightness level I selected."

> "So just for fun I tried turning on the fan, but any of the three fan speed buttons produced only high speed. 'What "smart" feature is this?', I wondered to myself. Maybe it needed to rest a while. So I turned off the fan & lights and went back to finish my dinner. For the rest of the evening the fan and lights would turn on and off

at random intervals and random levels, so I gave up on the idea that it would self-correct. So with a heavy heart I went over to the breaker panel, flipped the hood fan breaker to and fro, and the hood fan was once again well-behaved."

"For the next few days, my wife said that I was moping around as if someone had died. I would tell everyone I met, even complete strangers, about what happened: 'Hey, know what? I had to reboot my hood fan the other night!' The responses were varied, ranging from 'Freak!' to 'Sounds like what happened to my toaster...' Fellow programmers would either chuckle or stare in common disbelief."

"What's the embedded world coming to? Will programmers and companies everywhere realize the cost of their mistakes and clean up their act? Or will the entire world become accustomed to occasionally rebooting everything they own? Would the expensive embedded devices then come with a 'reset' button, advertised as a feature? Or will programmer jokes become as common and ruthless as lawyer jokes? I wish I knew the answer. I can only hope for the best, but I fear the worst."

One developer admitted to me that his consumer products company could care less about the correctness of firmware. Reboot—who cares? Customers are used to this, trained by decades of desktop computer disappointments. Hit the reset switch, cycle power, remove the batteries for 15 minutes, even pre-teens know the tricks of coping with legions of embedded devices.

Crummy firmware is the norm, but in my opinion is totally unacceptable. Shipping a defective product in any other field is like opening the door to torts. So far the embedded world has been mostly immune from predatory lawyers, but that Brigadoon-like isolation is unlikely to continue. Besides, it's simply unethical to produce junk.

But it's hard, even impossible, to produce perfect firmware. We must strive to make the code correct, but also design our systems to cleanly handle failures. In other words, a healthy dose of paranoia leads to better systems.

A Watchdog Timer is an important line of defense in making reliable products.

Well-designed watchdog timers fire off a lot, daily and quietly saving systems and lives without the esteem offered to other, human, heroes. Perhaps the developers producing such reliable WDTs deserve a parade. Poorly-designed WDTs fire off a lot, too, sometimes saving things, sometimes making them worse. A simple-minded watchdog implemented in a non-safety critical system won't threaten health or lives, but can result in systems that hang and do strange things that tick off our customers. No business can tolerate unhappy customers, so unless your code is perfect (whose is?) it's best in all but the most cost-sensitive apps to build a really great WDT.

An effective WDT is far more than a timer that drives reset. Such simplicity might have saved Clementine, but would it fire when the code tumbles into a really weird mode like that experienced by Ed's hood fan?

Internal WDTs

Internal watchdogs are those that are built into the processor chip. Virtually all highly integrated embedded processors include a wealth of peripherals, often with some sort of watchdog. Most are brain-dead WDTs suitable for only the lowest-end applications.

Let's look at a few. Toshiba's TMP96141AF is part of their TLCS-900 family of quite nice microprocessors, which offers a wide range of extremely versatile on-board peripherals. All have pretty much the same watchdog circuit. As the data sheet says, "The TMP96141AF is containing watchdog timer of Runaway detecting."

Ahem. And I thought the days of Jinglish were over. Anyway, the part generates a non-maskable interrupt when the watchdog times out, which is either a very, very bad idea or a wonderfully clever one. It's clever only if the system produces an NMI, waits a while, and only then asserts reset, which the Toshiba part unhappily cannot do. Reset and NMI are synchronous.

A nice feature is that it takes two different I/O operations to disable the WDT, so there are slim chances of a runaway program turning off this protective feature.

Motorola's widely-used 68332 variant of their CPU32 family (like most of these 68k embedded parts) also includes a watchdog. It's a simple-minded thing meant for low-reliability applications only. Unlike a lot of WDTs, user code must write two different values (0x55 and 0xaa) to the WDT control register to ensure the device does not time out. This is a very good thing—it limits the chances of rogue software accidentally issuing the command needed to appease the watchdog. I'm not thrilled with the fact that any amount of time may elapse between the two writes (up to the timeout period). Two back-to-back writes would further reduce the chances of random watchdog tickles, though once would have to ensure no interrupt could preempt the paired writes. And the 0x55/0xaa twosome is often used in RAM tests; since the 68k I/O registers are memory mapped, a runaway RAM test could keep the device from resetting.

The 68332's WDT drives reset, not some exception handling interrupt or NMI. This makes a lot of sense, since any software failure that causes the stack pointer to go odd will crash the code, and a further exception-handling interrupt of any sort would drive the part into a "double bus fault." The hardware is such that it takes a reset to exit this condition.

Motorola's popular Coldfire parts are similar. The MCF5204, for instance, will let the code write to the WDT control registers only once. Cool! Crashing code, which might do all sorts of silly things, cannot reprogram the protective mechanism. However, it's possible to change the reset interrupt vector at any time, pretty much invalidating the clever write-once design.

Like the CPU32 parts, a 0x55/0xaa sequence keeps the WDT from timing out, and back-to-back writes aren't required. The Coldfire datasheet touts this as an advantage since it can handle interrupts between the two tickle instructions, but I'd prefer less of a window. The Coldfire has a fault-on-fault condition much like the CPU32's double bus fault, so reset is also the only option when WDT fires—which is a good thing.

There's no external indication that the WDT timed out, perhaps to save pins. That means your hardware/software must be designed so at a warm boot the code can issue a from-the-ground-up reset to every peripheral to clear weird modes that may accompany a WDT timeout.

Philip's XA processors require two sequential writes of 0xa5 and 0x5a to the WDT. But like the Coldfire there's no external indication of a timeout, and it appears the watchdog reset isn't even a complete CPU restart—the docs suggest it's just a reload of the program counter. Yikes—what if the processor's internal states were in disarray from code running amok or a hardware glitch?

Dallas Semiconductor's DS80C320, an 8051 variant, has a very powerful WDT circuit that generates a special watchdog interrupt 128 cycles before automatically—and irrevocably—performing a hardware reset. This gives your code a chance to safe the system, and leave debugging breadcrumbs behind before a complete system restart begins. Pretty cool.

Summary: What's Wrong With Many Internal WDTs:

- A watchdog timeout must assert a hardware reset to guarantee the processor comes back to life. Reloading the program counter may not properly reinitialize the CPU's internals.

- WDTs that issue NMI without a reset may not properly reset a crashed system.

- A WDT that takes a simple toggle of an I/O line isn't very safe

- When a pair of tickles uses common values like 0x55 and 0xaa, other routines—like a RAM test—may accidentally service the WDT.

- Watch out for WDTs whose control registers can be reprogrammed as the system runs; crashed code could disable the watchdog.

- If a WDT timeout does not assert a pin on the processor, you'll have to add hardware to reset every peripheral after a timeout. Otherwise, though the CPU is back to normal, a confused I/O device may keep the system from running properly.

External WDTs

Many of the supervisory chips we buy to manage a processor's reset line include built-in WDTs.

TI's UCC3946 is one of many nice power supervisor parts that does an excellent job of driving reset only when Vcc is legal. In a nice small 8 pin SMT package it eats practically no PCB real estate. It's not connected to the CPU's clock, so the WDT will output a reset to the hardware safe-ing mechanisms even if there's a crystal failure. But it's too darn simple: to avoid a timeout just wiggle the input bit once in a while. Crashed code could do this in any of a million ways.

TI isn't the only purveyor of simplistic WDTs. Maxim's MAX823 and many other versions are similar. The catalogs of a dozen other vendors list equally dull and ineffective watchdogs.

But both TI and Maxim do offer more sophisticated devices. Consider TI's TPS3813 and Maxim's MAX6323. Both are "Window Watchdogs." Unlike the internal versions described above that avoid timeouts using two different data writes (like a 0x55 and then 0xaa), these require tickling within certain time bands. Toggle the WDT input too slowly, too fast, or not at all, and a timeout will occur. That greatly reduces the chances that a program run amok will create the precise timing needed to satisfy the watchdog. Since a crashed program will likely speed up or bog down if it does anything at all, errant strobing of the tickle bit will almost certainly be outside the time band required.

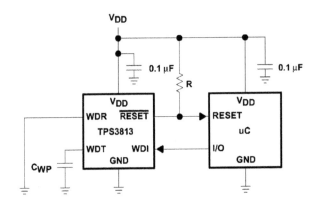

Figure 26-1: TI's TPS3813 is easy to use and offers a nice windowing WDT feature.

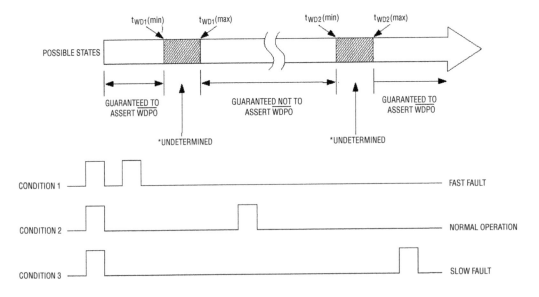

*UNDETERMINED STATES MAY OR MAY NOT GENERATE A FAULT CONDITION.

Figure 26-2: Window timing of Maxim's equally cool MAX6323.

Characteristics of Great WDTs

What's the rationale behind an awesome watchdog timer? The perfect WDT should detect all erratic and insane software modes. It must not make any assumptions about the condition of the software *or* the hardware; in the real world anything that can go wrong will. It must bring the system back to normal operation no matter what went wrong, whether from a software defect, RAM glitch, or bit flip from cosmic rays.

It's impossible to recover from a hardware failure that keeps the computer from running properly, but at the least the WDT must put the system into a safe state. Finally, it should leave breadcrumbs behind, generating debug information for the developers. After all, a watchdog timeout is the yin and yang of an embedded system. It saves the system, keeping customer happy, yet demonstrates an inherent design flaw that should be addressed. Without debug info, troubleshooting these infrequent and erratic events is close to impossible.

What does this mean in practice?

An effective watchdog is independent from the main system. Though all WDTs are a blend of interacting hardware and software, something external to the processor must always be poised, like the sword of Damocles, ready to intervene as soon as a crash occurs. Pure software implementations are simply not reliable.

There's only one kind of intervention that's effective: an immediate reset to the processor and all connected peripherals. Many embedded systems have a watchdog that initiates a non-maskable interrupt. Designers figure that firing off NMI rather than reset preserves some of the system's context. It's easy to seed debugging assets in the NMI handler (like a stack capture) to aid in resolving the crash's root cause. That's a great idea, except that it does not work.

All we really know when the WDT fires is that something truly awful happened. Software bug? Perhaps. Hardware glitch? Also possible. Can you ensure that the error wasn't some-thing that totally scrambled the processor's internal logic states? I worked with one system where a motor in another room induced so much EMF that our instrument sometimes went bonkers. We tracked this down to a sub-nanosecond glitch on one CPU input, a glitch so short that the processor went into an undocumented weird mode. Only a reset brought it back to life.

Some CPUs, notably the 68k and ColdFire, will throw an exception if a software crash causes the stack pointer to go odd. That's not bad… except that any watchdog circuit that then drives the CPU's non-maskable interrupt will unavoidably invoke code that pushes the system's context, creating a second stack fault. The CPU halts, staying halted till a reset, and only a reset, comes along.

Drive reset; it's the only reliable way to bring a confused microprocessor back to lucidity. Some clever designers, though, build circuits that drive NMI first, and then after a short delay pound on reset. If the NMI works then its exception handler can log debug information and then halt. It may also signal other connected devices that this unit is going offline for a while. The pending reset guarantees an utterly clean restart of the code. Don't be tempted to use the

NMI handler to safe dangerous hardware; that task always, in every system, belongs to a circuit external to the possibly confused CPU.

Don't forget to reset the whole computer system; a simple CPU restart may not be enough. Are the peripherals absolutely, positively, in a sane mode? Maybe not. Runaway code may have issued all sorts of I/O instructions that placed complex devices in insane modes. Give every peripheral a *hardware* reset; software resets may get lost in all of the I/O chatter.

Consider what the system must do to be totally safe after a failure. Maybe a pacemaker needs to reboot in a heartbeat (so to speak)… or maybe backup hardware should issue a few ticks if reboots are slow.

One thickness gauge that beams high energy gamma rays through 4 inches of hot steel failed in a spectacular way. Defective hardware crashed the code. The WDT properly closed the protective lead shutter, blocking off the 5 curie cesium source. I was present, and watched incredulously as the engineering VP put his head in path of the beam; the crashed code, still executing something, tricked the watchdog into opening the shutter, beaming high intensity radiation through the veep's forehead. I wonder to this day what eventually became of the man.

A really effective watchdog cannot use the CPU's clock, which may fail. A bad solder joint on the crystal, poor design that doesn't work well over temperature extremes, or numerous other problems can shut down the oscillator. This suggests that no WDT internal to the CPU is really safe. All (that I know of) share the processor's clock.

Under no circumstances should the software be able to reprogram the WDT or any of its necessary components (like reset vectors, I/O pins used by the watchdog, etc). Assume runaway code runs under the guidance of a malevolent deity.

Build a watchdog that monitors the *entire system's* operation. Don't assume that things are fine just because some loop or ISR runs often enough to tickle the WDT. A software-only watchdog should look at a variety of parameters to insure the product is healthy, kicking the dog only if everything is OK. What is a software crash, after all? Occasionally the system executes a HALT and stops, but more often the code vectors off to a random location, continuing to run instructions. Maybe only one task crashed. Perhaps only one is still alive—no doubt that which kicks the dog.

Think about what can go wrong in your system. Take corrective action when that's possible, but initiate a reset when it's not. For instance, can your system recover from exceptions like floating point overflows or divides by zero? If not, these conditions may well signal the early stages of a crash. Either handle these competently or initiate a WDT timeout. For the cost of a handful of lines of code you may keep a 60 Minutes camera crew from appearing at your door.

It's a good idea to flash an LED or otherwise indicate that the WDT kicked. A lot of devices automatically recover from timeouts; they quickly come back to life with the customer totally unaware a crash occurred. Unless you have a debug LED, how do you know if your precious creation is working properly, or occasionally invisibly resetting? One outfit complained that over time, and with several thousand units in the field, their product's response time to user inputs degraded noticeably. A bit of research showed that their system's watchdog properly drove the CPU's reset signal, and the code then recognized a warm boot, going directly to the application with no indication to the users that the time-out had occurred. We tracked the problem down to a floating input on the CPU, that caused the software to crash—up to several thousand times per second. The processor was spending most of its time resetting, leading to apparently slow user response. An LED would have shown the problem during debug, long before customers started yelling.

Everyone knows we should include a jumper to disable the WDT during debugging. But few folks think this through. The jumper should be *inserted to enable debugging*, and removed for normal operation. Otherwise if manufacturing forgets to install the jumper, or if it falls out during shipment, the WDT won't function. And there's no production test to check the watchdog's operation.

Design the logic so the jumper disconnects the WDT from the reset line (possibly though an inverter so an inserted jumper sets debug mode). Then the watchdog continues to function even while debugging the system. It won't reset the processor but will flash the LED. The light will blink a lot when breakpointing and singlestepping, but should never come on during full-speed testing.

Characteristics of Great WDTs:

- Make no assumptions about the state of the system after a WDT reset; hardware and software may be confused.

- Have *hardware* put the system into a safe state.

- Issue a hardware reset on timeout.

- Reset the peripherals as well.

- Ensure a rogue program cannot reprogram WDT control registers.

- Leave debugging breadcrumbs behind.

- *Insert* a jumper to disable the WDT for debugging; *remove* it for production units.

Using an Internal WDT

Most embedded processors that include high integration peripherals have some sort of built-in WDT. Avoid these except in the most cost-sensitive or benign systems. Internal units offer minimal protection from rogue code. Runaway software may reprogram the WDT controller, many internal watchdogs will not generate a proper reset, and any failure of the processor will make it impossible to put the hardware into a safe state. A great WDT must be independent of the CPU it's trying to protect.

However, in systems that really must use the internal versions, there's plenty we can do to make them more reliable. The conventional model of kicking a simple timer at erratic intervals is too easily spoofed by runaway code.

A pair of design rules leads to decent WDTs: kick the dog only after your code has done *several unrelated* good things, and make sure that erratic execution streams that wander into your watchdog routine won't issue incorrect tickles.

This is a great place to use a simple state machine. Suppose we define a global variable named "state." At the beginning of the main loop set state to 0x5555. Call watchdog routine A, which adds an offset—say 0x1111—to state and then ensures the variable is now 0x66bb. Return if the compare matches; otherwise halt or take other action that will cause the WDT to fire.

Later, maybe at the end of the main loop, add another offset to state, say 0x2222. Call watchdog routine B, which makes sure state is now 0x8888. Set state to zero. Kick the dog if the compare worked. Return. Halt otherwise.

This is a trivial bit of code, but now runaway code that stumbles into any of the tickling routines cannot errantly kick the dog. Further, no tickles will occur unless the entire main loop executes in the proper sequence. If the code just calls routine B repeatedly, no tickles will occur because it sets state to zero before exiting.

Add additional intermediate states as your paranoia or fear of litigation dictates.

Normally I detest global variables, but this is a perfect application. Cruddy code that mucks with the variable, errant tasks doing strange things, or any error that steps on the global will make the WDT timeout.

Do put these actions in the program's main loop, not inside an ISR. It's fun to watch a multitasking product crash—the entire system might be hung, but one task still responds to interrupts. If your watchdog tickler stays alive as the world collapses around the rest of the code, then the watchdog serves no useful purpose.

If the WDT doesn't generate an external reset pulse (some processors handle the restart internally) make sure the code issues a hardware reset to all peripherals immediately after start-up. That may mean working with the EEs so an output bit resets every resettable peripheral.

If you must take action to safe dangerous hardware, well, since there's no way to guarantee the code will come back to life, stay away from internal watchdogs. Broken hardware will obviously cause this… but so can lousy code. A digital camera was recalled recently when users found that turning the device off when in a certain mode meant it could never be turned on again. The code wrote faulty info to flash memory that created a permanent crash.

```
main(){
    state=0x5555;
    wdt_a();
    .
    .
    .
    .
    state+=0x2222;
    wdt_b();
}
```

```
wdt_a(){
    if (state!= 0x5555) halt;
    state+=0x1111;
}
```

```
wdt_b(){
    if (state!= 0x8888) halt;
    state=0;
    kick dog;
}
```

Figure 26-3: Pseudocode for handling an internal WDT.

An External WDT

The best watchdog is one that doesn't rely on the processor or its software. It's external to the CPU, shares no resources, and is utterly simple, thus devoid of latent defects.

Use a PIC, a Z8, or other similar dirt-cheap processor as a system health monitor. These parts have an independent clock, on-board memory, and the built-in timers we need to build a truly great WDT. Being external, you can connect an output to hardware interlocks that put dangerous machinery into safe states.

But when selecting a watchdog CPU check the part's specs carefully. Tying the tickle to the watchdog CPU's interrupt input, for instance, may not work reliably. A slow part—like most PICs—may not respond to a tickle of short duration. Consider TI's MSP430 family or processors. They're a very inexpensive (half a buck or so) series of 16 bit processors that use virtually no power and no PCB real estate.

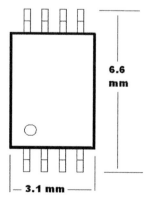

6.6 mm

3.1 mm

Figure 26-4: The MSP430—a 16 bit processor that uses no PCB real estate.
For metrically-challenged readers, this is about 1/4" x 1/8".

Tickle it using the same sort of state-machine described above. Like the windowed watch-dogs (TI's TPS3813 and Maxim's MAX6323), define min *and* max tickle intervals, to further limit the chances that a runaway program deludes the WDT into avoiding a reset.

Perhaps it seems extreme to add an entire computer just for the sake of a decent watchdog. We'd be fools to add extra hardware to a highly cost-constrained product. Most of us, though, build lower volume higher margin systems. A fifty cent part that prevents the loss of an expensive mission, or that even saves the cost of one customer support call, might make a lot of sense.

In a multiprocessor system it's easy to turn all of the processors into watchdogs. Have them exchange "I'm OK" messages periodically. The receiver resets the transmitter if it stops speaking. This approach checks a lot of hardware and software, and requires little circuitry.

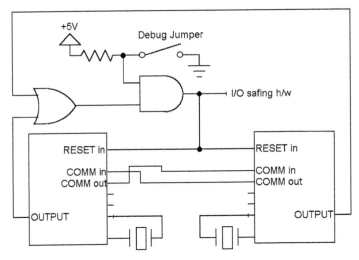

Figure 26-5: Watchdog for a dual-processor system—each CPU watches the other.

WDTs for Multitasking

Tasking turns a linear bit of software into a multidimensional mix of tasks competing for processor time. Each runs more or less independently of the others... which means each can crash on its own, without bringing the entire system to its knees.

You can learn a lot about a system's design just be observing its operation. Consider a simple instrument with a display and various buttons. Press a button and hold it down; if the display continues to update, odds are the system multitasks.

Yet in the same system a software crash might go undetected by conventional watchdog strategies. If the display or keyboard tasks die, the main line code or a WDT task may continue to run.

Any system that uses an ISR or a special task to tickle the watchdog but that does not examine the health of all other tasks is not robust. Success lies in weaving the watchdog into the fabric of all of the system's tasks—which is happily much easier than it sounds.

First, build a watchdog task. It's the only part of the software allowed to tickle the WDT. If your system has an MMU, mask off all I/O accesses to the WDT except those from this task, so rogue code traps on an errant attempt to output to the watchdog.

Next, create a data structure that has one entry per task, with each entry being just an integer.

When a task starts it increments its entry in the structure. Tasks that only start once and stay active forever can increment the appropriate value each time through their main loops.

Increment the data *atomically*—in a way that cannot be interrupted with the data half-changed. $++TASK_i$ (if TASK is an integer array) on an 8 bit CPU might not be atomic, though it's almost certainly OK on a 16 or 32 bitter. The safest way to both encapsulate and ensure atomic access to the data structure is to hide it behind another task. Use a semaphore to eliminate concurrent shared accesses. Send increment messages to the task, using the RTOS's messaging resources.

As the program runs the number of counts for each task advances. Infrequently but at regular intervals the watchdog task runs. Perhaps once a second, or maybe once a msec—it's all a function of your paranoia and the implications of a failure.

The watchdog task scans the structure, checking that the count stored for each task is reasonable. One that runs often should have a high count; another which executes infrequently will produce a smaller value. Part of the trick is determining what's reasonable for each task; stick with me—we'll look at that shortly.

If the counts are unreasonable, halt and let the watchdog timeout. If everything is OK, set all of the counts to zero and exit.

Why is this robust? Obviously, the watchdog monitors every task in the system. But it's also impossible for code that's running amok to stumble into the WDT task and errantly tickle the dog; by zeroing the array we guarantee it's in a "bad" state.

I skipped over a critical step—how do we decide what's a reasonable count for each task? It might be possible to determine this analytically. If the WDT task runs once a second, and one of the monitored tasks starts every 50 msec, then surely a count of around 20 is reasonable.

Other activities are much harder to ascertain. What about a task that responds to asynchronous inputs from other computers, say data packets that come at irregular intervals? Even in cases of periodic events, if these drive a low-priority task they may be suspended for rather long intervals by higher-priority problems.

The solution is to broaden the data structure that maintains count information. Add min and max fields to each entry. Each task must run at least min, but no more than max times.

Now redesign the watchdog task to run in one of two modes. The first is the one already described, and is used during normal system operation.

The second mode is a debug environment enabled by a compile-time switch that collects min and max data. Each time the WDT task runs it looks at the incremented counts and sets new min and max values as needed. It tickles the watchdog each time it executes.

Run the product's full test suite with this mode enabled. Maybe the system needs to operate for a day or a week to get a decent profile of the min/max values. When you're satisfied that the tests are representative of the system's real operation, manually examine the collected data and adjust the parameters as seems necessary to give adequate margins to the data.

What a pain! But by taking this step you'll get a great watchdog—and a deep look into your system's timing. I've observed that few developers have much sense of how their creations perform in the time domain. "It seems to work" tells us little. Looking at the data acquired by this profiling, though might tell a lot. Is it a surprise that task A runs 400 times a second? That might explain a previously-unknown performance bottleneck.

In a real time system we must manage and measure time; it's every bit as important as procedural issues, yet is oft ignored till a nagging problem turns into an unacceptable symptom. This watchdog scheme forces you to think in the time domain, and by its nature profiles—admittedly with coarse granularity—the time-operation of your system.

There's yet one more kink, though. Some tasks run so infrequently or erratically that any sort of automated profiling will fail. A watchdog that runs once a second will miss tasks that start only hourly. It's not unreasonable to exclude these from watchdog monitoring. Or, we can add a bit of complexity to the code to initiate a watchdog timeout if, say, the slow tasks don't start even after a number of hours elapse.

Summary and Other Thoughts

I remain troubled by the fan failure described earlier. It's easy to dismiss this as a glitch, an unexplained failure caused by a hardware or software bug, cosmic rays, or meddling by aliens. But others have written about identical situations with their vent fans, all apparently made by the same vendor.

When we blow off a failure, calling it a "glitch" as if that name explains something, we're basically professing our ignorance. There are no glitches in our macroscopically deterministic world. Things happen for a reason.

The fan failures didn't make the evening news and hurt no one. So why worry?

Surely the customers were irritated, and the possible future sales of that company at least somewhat diminished. The company escalated the general rudeness level of the world, and thus the sum total incipient anger level, by treating their customers with contempt. Maybe a couple more Valiums were popped, a few spouses yelled at, some kids cowered till dad calmed down. In the grand scheme of things perhaps these are insignificant blips. Yet we must remember the purpose of embedded control is to help people, to improve lives, not to help therapists garner new patients.

What concerns me is that if we cannot even build reliable fan controllers, what hope is there for more mission-critical applications?

I don't know what went wrong with those fan controllers, and I have no idea if a WDT—well designed or not—is part of the system. I do know, though, that the failures are unacceptable and avoidable. But maybe not avoidable by the use of a conventional watchdog. A WDT tells us the code is running. A windowing WDT tells us it's running with pretty much the right timing. No watchdog, though, flags software executing with corrupt data structures, unless the data is so bad it grossly affects the execution stream.

Why would a data structure become corrupt? Bugs, surely. Strange conditions the designers never anticipated will also create problems, like the never-ending flood of buffer overflow conditions that plague the net, or unexpected user inputs ("We never thought the user would press all 4 buttons at the same time!").

Is another layer of self-defense, beyond watchdogs, wise? Safety critical apps, where the cost of a failure is frighteningly high, should definitely include integrity checks on the data. Low threat equipment—like this oven fan—can and should have at least a minimal amount of code for trapping possible failure conditions.

Some might argue it makes no sense to "waste" time writing defensive code for a dumb fan application. Yet the simpler the system, the easier and quicker it is to plug in a bit of code to look for program and data errors.

Very simple systems tend to translate inputs to outputs. Their primary data structures are the I/O ports. Often several unrelated output bits get multiplexed to a single port. To change one bit means either reading the port's current status, or maintaining a copy of the port in RAM. Both approaches are problematic.

Computers are deterministic, so it's reasonable to expect that, in the absence of bugs, they'll produce correct results all the time. So it's apparently safe to read a port's current status, AND off the unwanted bits, OR in new ones, and output the result. This is a state machine; the outputs evolve over time to deal with changing inputs. But the process works only if the state machine never incorrectly flips a bit. Unfortunately, output ports are connected to the hostile environment of the real world. It's entirely possible that a bit of energy from starting the fan's highly inductive motor will alter the port's setting. I've seen this happen many times.

So maybe it's more reliable to maintain a memory image of the port. The downside is that a program bug might corrupt the image. Most of the time these are stored as global variables, so any bit of sloppy code can accidentally trash the location. Encapsulation solves that problem, but not the one of a wandering pointer walking over the data, or of a latent reentrancy issue corrupting things. You might argue that writing correct code means we shouldn't worry about a location changing, but we added a WDT to, in part, deal with bugs. Similar concerns about our data are warranted.

In a simple system look for a design that resets data structures from time to time. In the case of the oven fan, whenever the user selects a fan speed reset all I/O ports and data structures. It's that simple.

In a more complicated system the best approach is the oldest trick in software engineering: check the parameters passed to functions for reasonableness. In the embedded world we chose not to do this for three reasons: speed, memory costs, and laziness. Of these, the third reason is the real culprit most of the time.

Cycling power is the oldest fix in the book; it usually means there's a lurking bug and a poor WDT implementation. Embedded developer Peter Putnam wrote:

> Last November, I was sitting in one of a major airline's newer 737-900 aircraft on the ramp in Cancun, Mexico, waiting for departure when the pilot announced there would be a delay due to a computer problem. About twenty minutes later a group of maintenance personnel arrived. They poked around for a bit, apparently to no avail, as the captain made another announcement. "Ladies and Gentlemen," he said, "we're unable to solve the problem, so we're going to try turning off all aircraft power for thirty seconds and see if that fixes it."
>
> Sure enough, after rebooting the Boeing 737, the captain announced that "All systems are up and running properly."
>
> Nobody saw fit to leave the aircraft at that point, but I certainly considered it.

ASCII Table

Dec	Hex	Char	Dec	Hex	Char	Dec	Hex	Char	Dec	Hex	Char
0	00	Null	32	20	Space	64	40	@	96	60	`
1	01	Start of heading	33	21	!	65	41	A	97	61	a
2	02	Start of text	34	22	"	66	42	B	98	62	b
3	03	End of text	35	23	#	67	43	C	99	63	c
4	04	End of transmit	36	24	$	68	44	D	100	64	d
5	05	Enquiry	37	25	%	69	45	E	101	65	e
6	06	Acknowledge	38	26	&	70	46	F	102	66	f
7	07	Audible bell	39	27	'	71	47	G	103	67	g
8	08	Backspace	40	28	(	72	48	H	104	68	h
9	09	Horizontal tab	41	29	)	73	49	I	105	69	i
10	0A	Line feed	42	2A	*	74	4A	J	106	6A	j
11	0B	Vertical tab	43	2B	+	75	4B	K	107	6B	k
12	0C	Form feed	44	2C	,	76	4C	L	108	6C	l
13	0D	Carriage return	45	2D	-	77	4D	M	109	6D	m
14	0E	Shift out	46	2E	.	78	4E	N	110	6E	n
15	0F	Shift in	47	2F	/	79	4F	O	111	6F	o
16	10	Data link escape	48	30	0	80	50	P	112	70	p
17	11	Device control 1	49	31	1	81	51	Q	113	71	q
18	12	Device control 2	50	32	2	82	52	R	114	72	r
19	13	Device control 3	51	33	3	83	53	S	115	73	s
20	14	Device control 4	52	34	4	84	54	T	116	74	t
21	15	Neg. acknowledge	53	35	5	85	55	U	117	75	u
22	16	Synchronous idle	54	36	6	86	56	V	118	76	v
23	17	End trans. block	55	37	7	87	57	W	119	77	w
24	18	Cancel	56	38	8	88	58	X	120	78	x
25	19	End of medium	57	39	9	89	59	Y	121	79	y
26	1A	Substitution	58	3A	:	90	5A	Z	122	7A	z
27	1B	Escape	59	3B	;	91	5B	[	123	7B	{
28	1C	File separator	60	3C	<	92	5C	\	124	7C	|
29	1D	Group separator	61	3D	=	93	5D	]	125	7D	}
30	1E	Record separator	62	3E	>	94	5E	^	126	7E	~
31	1F	Unit separator	63	3F	?	95	5F	_	127	7F	▯

Dec	Hex	Char	Dec	Hex	Char	Dec	Hex	Char	Dec	Hex	Char
128	80	Ç	160	A0	á	192	C0	└	224	E0	α
129	81	ü	161	A1	í	193	C1	┴	225	E1	ß
130	82	é	162	A2	ó	194	C2	┬	226	E2	Γ
131	83	â	163	A3	ú	195	C3	├	227	E3	π
132	84	ä	164	A4	ñ	196	C4	─	228	E4	Σ
133	85	à	165	A5	Ñ	197	C5	┼	229	E5	σ
134	86	å	166	A6	ª	198	C6	╞	230	E6	µ
135	87	ç	167	A7	º	199	C7	╟	231	E7	τ
136	88	ê	168	A8	¿	200	C8	╚	232	E8	Φ
137	89	ë	169	A9	⌐	201	C9	╔	233	E9	⊙
138	8A	è	170	AA	¬	202	CA	╩	234	EA	Ω
139	8B	ï	171	AB	½	203	CB	╦	235	EB	δ
140	8C	î	172	AC	¼	204	CC	╠	236	EC	∞
141	8D	ì	173	AD	¡	205	CD	═	237	ED	ø
142	8E	Ä	174	AE	«	206	CE	╬	238	EE	ε
143	8F	Å	175	AF	»	207	CF	╧	239	EF	∩
144	90	É	176	B0	░	208	D0	╨	240	F0	≡
145	91	æ	177	B1	▒	209	D1	╤	241	F1	±
146	92	Æ	178	B2	▓	210	D2	╥	242	F2	≥
147	93	ô	179	B3	│	211	D3	╙	243	F3	≤
148	94	ö	180	B4	┤	212	D4	╘	244	F4	⌠
149	95	ò	181	B5	╡	213	D5	╒	245	F5	⌡
150	96	û	182	B6	╢	214	D6	╓	246	F6	÷
151	97	ù	183	B7	╖	215	D7	╫	247	F7	≈
152	98	ÿ	184	B8	╕	216	D8	╪	248	F8	°
153	99	Ö	185	B9	╣	217	D9	┘	249	F9	∙
154	9A	Ü	186	BA	║	218	DA	┌	250	FA	·
155	9B	¢	187	BB	╗	219	DB	█	251	FB	√
156	9C	£	188	BC	╝	220	DC	▄	252	FC	ⁿ
157	9D	¥	189	BD	╜	221	DD	▌	253	FD	²
158	9E	₧	190	BE	╛	222	DE	▐	254	FE	■
159	9F	ƒ	191	BF	┐	223	DF	▀	255	FF	□

Byte Alignment and Ordering

Sandeep Ahluwalia

Sandeep Ahluwalia is a writer and developer for EventHelix.com Inc., a company that is dedicated to developing tools and techniques for real-time and embedded systems. They are developers of the EventStudio, a CASE tool for distributed system design in object oriented as well as structured development environments. They are based in Gaithersburg, Maryland. Their website is www.eventhelix.com

Real-time systems consist of multiple processors communicating with each other via messages. For message communication to work correctly, the message formats should be defined unambiguously. In many systems this is achieved simply by defining C/C++ structures to implement the message format. Using C/C++ structures is a simple approach, but it has its own pitfalls. The problem is that different processors/compilers might define the same structure differently, thus causing incompatibility in the interface definition.

There are two reasons for these incompatibilities:
- byte alignment restrictions
- byte ordering

Byte Alignment Restrictions

Most 16-bit and 32-bit processors do not allow words and long words to be stored at any offset. For example, the Motorola 68000 does not allow a 16-bit word to be stored at an odd address. Attempting to write a 16-bit number at an odd address results in an exception.

Why Restrict Byte Alignment?

32-bit microprocessors typically organize memory as shown below. Memory is accessed by performing 32-bit bus cycles. 32-bit bus cycles can, however, be performed at addresses that are divisible by 4. (32-bit microprocessors do not use the address lines A1 and A0 for addressing memory.)

The reasons for not permitting misaligned long word reads and writes are not difficult to see. For example, an aligned long word X would be written as X0, X1, X2 and X3. Thus, the microprocessor can read the complete long word in a single bus cycle. If the same microprocessor now attempts to access a long word at address 0x000D, it will have to read bytes Y0, Y1, Y2 and Y3. Notice that this read cannot be performed in a single 32-bit bus cycle. The microprocessor will have to issue two different reads at address 0x100C and 0x1010 to read the complete long word. Thus, it takes twice the time to read a misaligned long word.

	Byte 0	Byte 1	Byte 2	Byte 3
0x1000				
0x1004	X0	X1	X2	X3
0x1008				
0x100C		Y0	Y1	Y2
0x1010	Y3			

Compiler Byte Padding

Compilers have to follow the byte alignment restrictions defined by the target microprocessors. This means that compilers have to add pad bytes into user defined structures so that the structure does not violate any restrictions imposed by the target microprocessor.

The compiler padding is illustrated in the following example. Here a char is assumed to be one byte, a short is two bytes and a long is four bytes.

User Defined Structure

```
struct Message
{
    short opcode;
    char subfield;
    long message_length;
    char version;
    short destination_processor;
};
```

Actual Structure Definition Used By the Compiler

```
struct Message
{
    short opcode;
    char subfield;
    char pad1          // Pad to start the long word at a 4 byte
                          boundary
    long message_length;
    char version;
    char pad2;         // Pad to start a short at a 2 byte boundary
    short destination_processor;
    char pad3[4];      // Pad to align the complete structure to a
                          16 byte boundary
};
```

In the above example, the compiler has added pad bytes to enforce byte alignment rules of the target processor. If the above message structure was used in a different compiler/microprocessor combination, the pads inserted by that compiler might be different. Thus two applications using the same structure definition header file might be incompatible with each other.

Thus it is a good practice to insert pad bytes explicitly in all C-structures that are shared in a interface between machines differing in either the compiler and/or microprocessor.

General Byte Alignment Rules

The following byte padding rules will generally work with most 32-bit processors. You should consult your compiler and microprocessor manuals to see if you can relax any of these rules.

- Single byte numbers can be aligned at any address
- Two byte numbers should be aligned to a two byte boundary
- Four byte numbers should be aligned to a four byte boundary
- Structures between 1 and 4 bytes of data should be padded so that the total structure is 4 bytes.
- Structures between 5 and 8 bytes of data should be padded so that the total structure is 8 bytes.
- Structures between 9 and 16 bytes of data should be padded so that the total structure is 16 bytes.
- Structures greater than 16 bytes should be padded to 16 byte boundary.

Structure Alignment for Efficiency

Sometimes array indexing efficiency can also determine the pad bytes in the structure. Note that compilers index into arrays by calculating the address of the indexed entry by the multiplying the index with the size of the structure. This number is then added to the array base address to obtain the final address. Since this operation involves a multiply, indexing into arrays can be expensive. The array indexing can be considerably speeded up by just making sure that the structure size is a power of 2. The compiler can then replace the multiply with a simple shift operation.

Byte Ordering

Microprocessors support big-endian and little-endian byte ordering. Big-endian is an order in which the "big end" (most significant byte) is stored first (at the lowest address). Little-endian is an order in which the "little end" (least significant byte) is stored first.

The table below shows the representation of the hexadecimal number 0x0AC0FFEE on a big-endian and little-endian machine. The contents of memory locations 0x1000 to 0x1003 are shown.

	0x1000	0x1001	0x1002	0x1003
Big-endian	0x0A	0xC0	0xFF	0xEE
Little-endian	0xEE	0xFF	0xC0	0x0A

Why Different Byte Ordering?

This is a difficult question. There is no logical reason why different microprocessor vendors decided to use different ordering schemes. Most of the reasons are historical. For example, Intel processors have traditionally been little-endian. Motorola processors have always been big-endian.

The situation is quite similar to that of Lilliputians in Gulliver's Travels. Lilliputians were divided into two groups based on the end from which the egg should be broken. The big-endians preferred to break their eggs from the larger end. The little-endians broke their eggs from the smaller end.

Conversion Routines

Routines to convert between big-endian and little-endian formats are actually quite straight-forward. The routines shown below will convert from both ways—i.e., big-endian to little-endian and back.

Big-endian to Little-endian conversion and back

```
short convert_short(short in)
{
    short out;
    char *p_in = (char *) &in;
    char *p_out = (char *) &out;
    p_out[0] = p_in[1];
    p_out[1] = p_in[0];
    return out;
}

long convert_long(long in)
{
    long out;
    char *p_in = (char *) &in;
    char *p_out = (char *) &out;
    p_out[0] = p_in[3];
    p_out[1] = p_in[2];
    p_out[2] = p_in[1];
    p_out[3] = p_in[0];
    return out;
}
```

Index

A

AC circuits, 14
active devices, 20
adder, 36
Agilent, 50
alignment, 260
alternating current, 14
ampere, 5
AND gates, 37
aperiodic, 14
arccosine, 196
archive, 146
arcsine, 196
arctangent, 196
assert(), 275
asynchronous, 236
atomic variables, 231

B

banning interrupts, 246
Barry Boehm, 209
base, 21, 33
battery-backed up, 307
BCD, 36
BDM, 50
Bessel, 181
big bang, 78
big-endian, 162
binary, 33
binary coded decimal, 36
binary point, 167
bistable, 43
block-copied code, 71
Boolean, 42
Branch, 144

breadcrumbs, 345
breakpoint, 276
bridge, 25
brown-outs, 307
build, 144

C

callbacks, 323
CAN, 64
CANopen, 64
capacitive reactance, 15
capacitors, 15
change requests, 96
Chebyshev, 181
check lists, 92
check-in, 143
checkout, 143
checkpointing, 145
checksum, 300
circuits, 11
Clementine, 339
clip-on, 50
clock, 43
COCOMO, 209
code browser, 62
code compression, 257
code generator tools, 63
code inspections, 72
code reviews, 62
CodeTest, 277
coding, 33
cohesion, 70
coil, 19
Coldfire, 342
collector, 21

Printed and bound by CPI Group (UK) Ltd, Croydon, CR0 4YY

03/10/2024

01040338-0004